RAND NATIONAL DEFENSE RESEARCH INSTITUTE

T0108944

National Security Implications of Virtual Currency

Examining the Potential for Non-state Actor Deployment

Joshua Baron, Angela O'Mahony, David Manheim, Cynthia Dion-Schwarz

Prepared for the Office of the Secretary of Defense

Approved for public release; distribution unlimited

For more information on this publication, visit www.rand.org/t/rr1231

Library of Congress Cataloging-in-Publication Data
ISBN: 978-0-8330-9183-3

Published by the RAND Corporation, Santa Monica, Calif.
© Copyright 2015 RAND Corporation
RAND® is a registered trademark.

www.rand.org

Preface

This report examines the feasibility for non-state actors to increase their political and/or economic power by deploying a virtual currency (VC) for use in regular economic transactions. Bitcoin is a digital representation of valued currency that, like conventional currency, can be transferred, stored, or traded electronically. VCs are neither issued by a central bank or public authority nor are they necessarily linked to a fiat currency (dollars, euros, etc.). This form of currency is accepted by people as a means of payment. We addressed the following research questions from both the technological and political-economic perspectives:

- Why would a non-state actor deploy a VC? That is, what political and/or economic utility is there to gain? How might this non-state actor go about such a deployment? What challenges would it have to overcome?
- How might a government or organization successfully technologically disrupt a VC deployment by a non-state actor, and what degree of cyber sophistication would be required?
- What additional capabilities become possible when the technologies underlying the development and implementation of VCs are used for purposes broader than currency?

This report should be of interest to policymakers interested in technology, counterterrorism, and intelligence and law enforcement issues, as well as for VC and cybersecurity researchers.

This research was sponsored by the Office of the Secretary of Defense, and it was conducted within the International Security and

Defense Policy Center of the RAND National Defense Research Institute, a federally funded research and development center sponsored by the Office of the Secretary of Defense, the Joint Staff, the Unified Combatant Commands, the Navy, the Marine Corps, the defense agencies, and the defense Intelligence Community.

For more information on the RAND International Security and Defense Policy Center, see http://www.rand.org/nsrd/ndri/centers /isdp.html or contact the Center director (contact information provided on the web page).

Comments or questions about this report should be addressed to the project leader, Joshua Baron, at Joshua_Baron@rand.org.

Contents

Figures and Tables

Figures

Tables

Summary

A virtual currency (VC) is a digital representation of value that can be transferred, stored, or traded electronically and that is neither issued by a central bank or public authority, nor necessarily attached to a fiat currency (dollars, euros, etc.), but is accepted by people as a means of payment. Currently the most popular VC is Bitcoin. The national security–policy implications of the rise of virtual-currency technology is the subject of much debate as of late. There has been a particular focus on the potential anonymity of VCs such as Bitcoin as well as the potential for terrorist or insurgent group usage in a manner resilient against efforts by local and global law enforcement, military, and intelligence organizations (including those of the United States) to survey. The goal of this report is to enrich this policy conversation by providing an in-depth analysis of the technological issues associated with virtual currencies.

This report examines the potential for non-state actors, including terrorist and insurgent groups, to increase their political and/or economic power by *deploying a VC as a medium for regular economic transactions* as opposed to exploiting already-deployed virtual currencies, such as Bitcoin, as a means of illicit transfer, fundraising, or money laundering.

We examine the issue of VC deployment from both the technological and political-economic perspectives, with a particular focus on the challenges facing non-state actors who attempt a VC deployment. These challenges inform how the United States, its allies, and other cyber actors might respond to such a VC deployment if it threatened their national-security interests. To date, there has not been a case of such a non-state actor deployment; in this report, we aim to high-light those key issues that might serve as technological and political-economic barriers today in order to understand why such a deployment may become more feasible—and beneficial—for the non-state actor in the future.

We will also briefly examine the broader technological implications of virtual currencies and the availability of their derivative technologies to unsophisticated users in cyberspace. We first investigate technologies that development of VCs may advance, including a general increased sophistication in cryptographic applications. More generally, we make the case that the main technological contribution of decentralized virtual currencies, from a national-security perspective, is cyber resilience and ask: What would the policy implications be if unsophisticated cyber actors had persistent, assured access to cyber services *regardless* of whether a highly sophisticated state actor opposes their use?

Our key research questions and the answers we derived are as follows:

- Why would a non-state actor deploy a VC? That is, what political and/or economic utility would be realized? How might the actor go about such a deployment? What challenges would the actor have to overcome?
  - Deploying a VC may be an attractive alternative for non-state actors who look to disrupt sovereignty and increase their own political and/or economic power by displacing state-based currencies. VC deployments are particularly attractive in developing countries and in countries undergoing internal turmoil, where the existing financial infrastructure is either insufficient or weakened. The rapid deployment of a VC over a large geographic area would likely be less complicated than deploying more common currencies, such as those based on commodities or paper-based currencies. Examples of relevant non-state actors considered here include terrorist organizations, insurgent groups, drug cartels, and other criminal organizations.
  - The use of an established VC, such as Bitcoin, as a currency by a non-state actor would provide few political or economic

advantages and would likely be vulnerable to cyber attack by a sophisticated adversary, while facing many of the same implementation challenges as a new VC.

- Developing a VC from scratch, however, requires high technological sophistication, extensive networking and computational infrastructure, and enough expertise to ensure successful roll-out and adoption, all of which are in short supply among non-state actors. Specific challenges include developing the software for a capable, secure VC; deploying the means of physically transacting with a VC, particularly in countries with fewer smartphones; and overcoming the ability of nation-states to launch successful cyber attacks against a VC.

- From an economic perspective, promoting adoption of VCs (versus adopting established currencies) may face significant challenges of acceptance by the population in which the VC is implemented, both as a new currency with no previous history and thus potentially lacking in legitimacy and as a currency that is intangible in societies accustomed to conceiving of money in terms of its physical manifestations. We expect suspicions of VCs will erode, however, as they become more familiar with them. Changes in attitude can take place as the technologies that underlie VCs become more prevalent and trusted. Moreover, in a territory in which a VC is the only medium of exchange, economic necessity may force people to accept VCs where they would have otherwise rejected them.

- The deployment of a VC by non-state actors would be easier, and indeed are most feasible today, when supported by a nation-state with advanced cyber expertise. This nation-state could enable the non-state actors to overcome the considerable technical hurdles associated with deploying a VC. There are numerous parts of the world from which such support might originate, e.g., Iran (as in its support of Hezbollah and formerly of Hamas) or Russia (Ukrainian separatists).

- In spite of current hurdles, the trends indicate a future in which VCs could be deployed by non-state actors or other organizations, particularly given the rapid rate at which the needed technologies are becoming commodities available for purchase and the gradual but widening public understanding of VCs.

- How might a government or organization successfully disrupt a new VC deployment by a non-state actor and what degree of cyber sophistication would be required?
  - It would be difficult for a non-state actor to structurally design a VC that would be both resilient to attack and usable by all persons in the non-state actor's geographic area of influence. Such difficulty is especially exacerbated in less technologically sophisticated regions and in areas with incomplete networking infrastructures.
  - VCs are vulnerable to attacks of varying degrees of sophistication.
    - Relatively unsophisticated attacks by governments, other non-state actors, or even users of another VC could involve distributed denial-of-service attacks against more centralized services, such as mining pools or online-wallet applications, or attempts to gain control of a VC via exploiting a VC's market rules, e.g., by supplying a majority of the computing power for Bitcoin-like VCs.
    - A more sophisticated attacker could conduct zero-day exploits—attacks that take advantage of a software vulnerability that the developer is unaware of and for which no patch exists. Zero-day attacks could target VC services, such as exchanges and wallets, as well as cell-phone applications used for common transactions.
    - The most sophisticated challengers could attack the underlying VC infrastructure, including hardware, or covertly corrupt the software used by VC participants, including through the subversion of the underlying security mechanisms on which the software relies.
- What additional cyber capabilities other than VC use become possible, not just for non-state actors, as the technologies underlying the development and implementation of VCs continue to

mature? The further development and implementation of VCs could contribute to security-related technological developments outside the currency arena, which could aid non-state actors.

– VCs demonstrate a resilient means of storing data in a highly distributed fashion that is very hard to corrupt; possible implications of this include information dissemination (blogs, social media, forums, news websites) that is eventually completely resilient to nation-state interference.

– The need to develop security mechanisms for VCs could encourage the development of advanced cryptographic techniques, such as secure multiparty computation, which seeks to perform distributed computation while preserving the confidentiality of inputs and outputs in the presence of malicious activity.

– VCs represent the latest step toward decentralized cyber services. In particular, the historical trend suggests the development of a resilient public cyber key terrain, which this report defines as the ability of unsophisticated cyber actors to have persistent, assured access to cyber services regardless of whether a highly sophisticated state actor opposes their use. This has implications for national firewalls, access to extremist rhetoric, the feasibility of nation-state cyber attacks, and the ability to maintain uninterruptible and anonymous encrypted links.

The Department of Defense should be aware of the following: VCs are an increasingly technologically feasible tool for non-state actors to deploy; efforts to destabilize confidence in a new VC are effective, while popular sentiment is still untrusting of VCs for common transactions; VCs are just like any other service in cyberspace, and methods to successfully attack them are not meaningfully different than for any other cyberspace operation; decentralization affords more, though not total, resilience to disruptions from cyber attacks; and finally, the trend toward decentralized cyber service will only make it easier for unsophisticated cyber actors to have increasingly resilient access to cyber services, which is a two-way street that could enable unprecedented global access to information and communication services that, at its core, could be both beneficial and harmful to the national security interests of the United States.

Acknowledgments

This work would not have been possible without the spearheading and guidance of many RAND colleagues. We are particularly grateful to Ryan Henry, who pushed for this project to be undertaken in the first place. Thanks go to Seth Jones for his support and guidance, Michael McNerney for his work scoping the project at its outset, and Christopher Chivvis for his strong guidance on the draft of this work. We also thank Lillian Ablon, Lieutenant Colonel William Fry (U.S. Air Force), Richard Neu, Howard Shatz, and Cortney Weinbaum for sharing their considerable expertise.

Thanks go to Yasha Tabrizy (Department of the Treasury) and Ryan Otteson (formerly of the Federal Bureau of Investigation) for pro-viding important and valuable insight and context.

We are grateful to Aaron Brantly of the United States Military Academy at West Point, our colleague Krishna Kumar at RAND, and Tim Maurer of New America for their careful reviews of the manuscript.

We also thank Holly Johnson, James Chiesa, Erin Dick, Theresa DiMaggio, Christopher Dirks, and Nedda Rahme for their aid in editing this document.

That we received help and insights from those acknowledged should not be taken to imply that they concur with the views expressed in this report. Any errors are the authors' alone.

Abbreviations

DARPA	Defense Advanced Research Projects Agency
DDoS	distributed denial of service
DoD	Department of Defense
HUMINT	human intelligence
IP	Internet protocol
ISIL	Islamic State of Iraq and the Levant
MPC	secure multiparty computation
TTP	tactics, techniques, and procedure
VC	virtual currency
VOIP	voice over Internet protocol
ZK-SNARKs	zero-knowledge succinct arguments of knowledge

CHAPTER ONE

Introduction

With the introduction and growing conversation about Bitcoin, interest in virtual currencies (VCs) has dramatically increased.[1] This interest is diverse across many communities: from venture capitalists to cybersecurity academics to economists. In addition, organized groups and governments have explored or adopted VCs for a variety of legitimate and illegitimate purposes, albeit with mixed success. Today, the utility of VCs, both in the near and long term, remains the subject of intense debate.

A VC, when issued as a currency for everyday transactions, requires considerably less new physical infrastructure than government-backed currencies in broad use today. VCs, however, also require a networked architecture capable of supporting such everyday transactions. As a

---

[1] See Satoshi Nakamoto, 2008, for a discussion of Bitcoin. See also the section in this report "Origin and Trends of Virtual Currencies" in Chapter Two for further examples of historical and current VCs. The European Banking Authority ("EBA Opinion on 'Virtual Currencies,'" July 4, 2014) gives a working definition of a *virtual currency*: "a digital representation of value that is neither issued by a central bank or a public authority, nor necessarily attached to a [fiat currency], but is accepted by natural or legal persons as a means of payment and can be transferred, stored, or traded electronically." The requirement by the European Banking Authority that VCs are not issued by a central bank of a public authority will be the subject of some discussion in this report. For further discussion of the definition of a VC, see also European Central Bank, *Virtual Currency Schemes*, October 2012. Throughout this report, as a matter of convention, we will use the term *virtual currencies* or *VCs* rather than *digital currencies* or *cryptocurrencies*. It should be noted that not all VCs are cryptocurrencies, but by the definition used by this report, all cryptocurrencies are VCs. While a definitional distinction could be made between virtual and digital currencies, we will treat them as the same.

result, the rapid deployment of a VC over a large geographic area may be considerably less complicated than deploying these more traditional currencies; the amount of labor, capital, and infrastructure required to deploy a VC has the potential to be dramatically less. In developing countries and in countries undergoing internal turmoil where the existing financial infrastructure is either insufficient or weakened, and where legal enforcement is weak, deployment of VCs may be an attractive alternative for non-state actors seeking to disrupt sovereignty and increase their own political and/or economic power. Examples of relevant non-state actors considered here include terrorist organizations, insurgent groups, drug cartels, and criminal organizations.

The United States national-security community should understand how these non-state actors might exploit VCs as another tool to increase their influence in areas of interest to U.S. foreign and national-security policy in order to understand the threat as well as assess how the threat may be best thwarted. Accordingly, this report is mainly interested in how the United States or another opponent of a VC deployment can leverage or increase the challenges to deploying a VC for common transactions. This examination is a small part of a larger conversation on the feasibility of VCs, both from a social-science perspective (i.e., VC as currency) as well as a technological perspective (i.e., VC as secure, anonymous, and resilient cyber service).

This report will examine the potential for terrorist, insurgent, or criminal groups to increase their political and/or economic power by *deploying* a VC to use as a currency for regular economic transactions rather than *exploiting* existing VCs as a means of illicit transfer, fundraising, or money laundering. We have chosen to primarily examine VC deployment rather than exploitation for several reasons. First, while we are aware of literature that has examined the exploitation of VCs for these ends,[2] there is little literature that examines VC usage as a complete replacement currency to the indigenous or other common

---

[2]   For instance, see Raj Samani, "Cybercrime Exposed: Cybercrime-as-a-Service," corporate white paper, Santa Clara, Calif.: McAfee Labs, 2013a, and "Digital Laundry: An Analysis of Online Currencies, and Their Use in Cybercrime," corporate white paper, Santa Clara, Calif.: McAfee Labs, 2013b; and Aaron Brantly, "Financing Terror Bit by Bit," *CTC Sentinel*, Vol. 7, No. 10, October 2014, pp. 1–5.

currencies. Second, examining VC use in this setting will also increase understanding about VCs in general; were a non-state actor to deploy a VC in a country whose legitimate government objects to its usage, significant questions of security, anonymity, and resilience to cyber attack will arise. Answering these questions, particularly in the case where (possibly allied) nation-states with sophisticated cyber capabilities (such as the United States) may be involved in attempting to disrupt a non-state actor VC deployment, pushes our current understanding of VCs and is likely to have positive spillovers to the area of cybersecurity technologies. Finally, this examination will help reveal those key issues that critically enable or impede VC deployment, yielding implications for VC exploitation.

After providing some background on the evolution of VCs and their relevance to non-state actors (Chapter Two), this report proceeds along two lines of inquiry: political and economic, and technological. We will discuss how VCs can help in projecting political power (Chapter Three). We then examine how a non-state actor (especially terrorists or insurgents) might feasibly deploy a VC in developing or failed nation-states (Chapter Four). We also examine how VCs might be disrupted in both deployment and operations, either through deliberate actions by a third party or through implementation failures.

Finally (Chapter Five), we examine VCs within a larger technological perspective: What capabilities become possible when the technologies underlying VCs are used for different, broader purposes beyond currencies for economic transactions? In particular, we examine the implications of enabling low-sophistication cyber actors to have access to resilient cyber services that would otherwise only be available to actors with far greater sophistication.

Approach

This analysis is based on extensive literature reviews and interviews with subject-matter experts in both the technical aspects as well the usage of VCs. As much as possible, we relied on published academic literature, policy literature, white papers of established security organizations, and formal documentation of the various VCs. We avoided blog posts and websites when possible, due to their often-tenuous reliability; however, it is not possible to avoid them entirely, particularly in the dynamically changing world of VCs.[3]

In this paper, we call adversaries of the non-state actor deploying a VC *opponents*; these opponents may include both the nation-state(s) where a VC was deployed as well as allies of that "victim" nation-state, who may have far more advanced cyber capabilities (such as the United States).

---

[3] In particular, we refer to wikis for information about VCs numerous times, namely in regard to Bitcoin. This is done both because these sites are the best reference and ideally these websites will adjust over time to provide the most accurate, current picture of a dynamically changing VC environment. The drawback of such an approach is that some citations may be inaccessible or less well maintained some time after the publication of this report.

CHAPTER TWO

The Current State of Virtual Currencies

This chapter provides an introduction to VCs that we build on in the rest of the report. It may be of independent interest as a primer on VCs for the interested reader. We first examine the economic progression to VCs in order to understand them from a social-science perspective. We then examine the current technological state-of-the-art of VCs and introduce the main currencies, most notably Bitcoin. Finally, we briefly highlight current non-state actor use of VCs.

The Evolution to Virtual Currencies

We briefly examine here the historical evolution of currencies, from gold to VCs, to ascertain the reasons motivating a VC's use. As a motivation for this analysis, many currency users prefer transactions that are secure and anonymous; virtually all users prefer that the transactions take place within a system that is stable, resilient, and easy to use. Superficially, a decentralized VC such as Bitcoin appears far removed from the gold coins often used as a comparison. VCs have no physical manifestation, they have no intrinsic value, and their value is generally not backed by a government.

Gold coins have been used as a store of value, unit of account, and medium of exchange since at least 700 BCE.[1] As a currency, gold has

---

[1] See Peter L. Bernstein, *The Power of Gold: The History of an Obsession*, Hoboken, N.J.: Wiley and Sons, Inc., 2004, p. 24. These are the three canonical functions of money. Gold has been used in bar form as money for much longer.

many desirable properties.[2] It is a commodity that has a market value in and of itself (i.e., intrinsic value). As Peter Bernstein notes, however,

> Value alone is insufficient for a substance to qualify as money. Lots of things have value that do not serve as money. In fact, the most effective forms of money have developed from objects that were otherwise quite useless, such as paper and computer blips.[3]

Unlike the cowrie shells that were used as a trading currency in West Africa, gold is relatively indestructible.[4] The supply of gold in the world has been plentiful enough to sustain its use as a currency, but not so plentiful as to erode its value. This stands in contrast to other metals, such as platinum (too rare) and aluminum (too abundant). Gold is also easily divisible, making it easy to measure.

Although gold and silver coins may be issued by a government, their value lies primarily in their weight and purity. As a result, a central authority is not necessary for enforcing the value of a commodity-based currency. Commodity-based currencies are also highly anonymous. There is no record built into any of the transactions for which the currency is used that tracks users or uses. Although most commodity-based currencies have maintained stable values over time, commodity-based currencies have been vulnerable to value fluctuations beyond the control of any monetary authorities. This is because the value basis of the currency reflects the supply and demand for the commodity. For example, the value of silver vis-à-vis gold fell by one-half around 1870 as silver discoveries in Mexico, as well as reduced demand for silver as currency in Europe, increased the supply of silver, essentially decreasing the demand.[5] In addition to commodity-based currencies' vulnerabilities to value instability, they are difficult to use for anything more

---

[2]   Silver has similar properties. We focus on gold in this discussion for simplicity.

[3]   Bernstein, 2004.

[4]   See Marion Johnson, "The Cowrie Currencies of West Africa. Part I," *Journal of African History*, Vol. 11, No. 1, 1970, pp. 17–49.

[5]   Jeffry A. Frieden, *Global Capitalism: Its Fall and Rise in the Twentieth Century*, New York: W. W. Norton and Company, 2006.

than small, local transactions because they are physically inconvenient to transfer at scale and distance.

Over time, most countries migrated from commodity-based to paper (fiat) currencies; these currencies are decreed by a central authority to be legal tender, have no intrinsic value, and are only convertible into a commodity such as gold at the discretion of a central authority.[6] As a result, fiat currencies' value depends on users' trust that the central authority will be able to maintain the currency's value. Fiat currencies have key advantages over commodity-based currencies. They are lighter and easier to use (although still difficult to transport over distance), and they provide more leverage for governments to control monetary and fiscal policy. Similar to commodity-based currencies, fiat currencies can provide more anonymous transactions. Fiat currencies, however, are highly dependent on their central authority to maintain their value. The stability of fiat currencies is dependent on governments' macroeconomic policies and can experience huge fluctuations, even becoming worthless (e.g., during hyperinflationary episodes).

Financial innovations have allowed people to conduct economic transactions far beyond the constraints imposed by physical currency. Bills of exchange emerged around the great European trade fairs that took place in the 1200s to facilitate commerce without having to ship large quantities of gold from town to town and country to country.[7] These bills of exchange were denominated in countries' currencies, similar to the modern form of writing a check against money in a check-

---

[6] This very simplified evolution of monetary systems does not discuss alternatives to territorial monetary systems. For more detailed discussions, see Benjamin J. Cohen, *The Geography of Money*, Ithaca, N.Y.: Cornell University Press, 1998; Glyn Davies, *A History of Money: From Ancient Times to the Present Day*, Chicago: University of Chicago Press, 2005; Eric Helleiner, *The Making of National Money: Territorial Currencies in Historical Perspective*, Ithaca, N.Y.: Cornell University Press, 2003; and Jack McIver Weatherford, *The History of Money*, New York: Crown Publishers, 1997. An important step between the current monetary system and the pre–World War II commodity-based monetary system was the Bretton Woods monetary system (1944–1971), in which the U.S. dollar was backed by gold reserves, other developed country currencies were pegged to the dollar, and the developing countries' currencies were pegged to a basket of developed country currencies (Frieden, 2006).

[7] See Charles Kindleberger, *A Financial History of Western Europe*, Oxford: Oxford University Press, 1993.

ing account. More recent technological innovations have allowed users to move away from paper-based exchange systems (such as checks) to electronic systems (such as swiping debit cards through a point-of-sale card reader) to using near-field communication (NFC) technology to enable radio communication through mobile-computing platforms (such as via applications on smartphones).[8] As with the 13th-century bills of exchange, these innovations are convenient mechanisms that allow users to use traditional currencies more efficiently. Unlike VCs, they do not constitute new currencies.

VCs have become increasingly common in recent years. So far, no VCs are fiat currencies—no government has adopted a VC as its legal tender. They do, however, represent value for a particular community that uses them as a means of exchange. VCs have been used in online gaming communities and loyalty programs, such as airline frequent-flier programs, to keep track of redeemable membership credits that may not otherwise have value in terms of a fiat currency.[9] VCs, such as money used in online games or frequent-flier miles, are designed to act as a store of value, unit of account, and medium of exchange solely within their community of interest. That community of interest does not, however, need to occupy a single geographical or political unit.

Some of the latest VCs, such as Bitcoin, differ from earlier VCs in that they are designed explicitly to function as currency in the real economy and are exchangeable for government-issued fiat currencies. Returning to the comparison with gold coins, Bitcoin shares many of the same characteristics of gold coins. There is a limited supply of currency in circulation. Similar to a commodity such as gold, Bitcoin's exchange rate can be volatile. Bitcoin is easily measurable and divisible. In contrast to gold, Bitcoin is easily transportable and does not need to transit through international borders as currency, which may increase its ease of use and reduce cross-border transaction costs (as well as chal-

---

[8]   This is the technology underlying applications, such as Google Wallet, Apple Pay, and Venmo.

[9]   Exchanges may develop to allow users to "cash out" VCs for fiat currencies, but this is neither a feature nor a requirement of VCs.

lenge law enforcement and intelligence efforts). Finally, Bitcoin does not depend on a central authority to safeguard its value.

Perhaps the most important distinction between Bitcoin and previous VCs is that while VCs do not technically require a central authority, one of Bitcoin's key features is its completely *decentralized* authority—and many VCs have followed Bitcoin precisely in this direction. As a result, VCs such as Bitcoin cannot build trust in their currencies' stability based on the policies and capacities of a central authority. Instead, users' trust in VCs depends on their trust in the decentralized mechanisms that secure and sustain a VC. Current VCs have authority structures that range from completely centralized to completely decentralized (see Figure 2.1).

Having examined the evolution to VCs from a monetary perspective, we will now examine the evolution of the VCs themselves, mainly from a technological perspective.

Figure 2.1
Virtual Currencies Have Varied Authority Structures

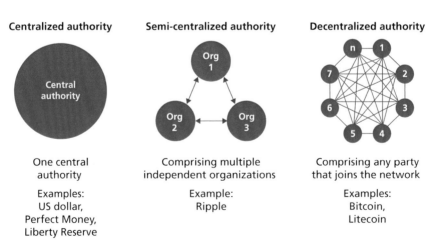

Centralized authority	Semi-centralized authority	Decentralized authority
One central authority	Comprising multiple independent organizations	Comprising any party that joins the network
Examples: US dollar, Perfect Money, Liberty Reserve	Example: Ripple	Examples: Bitcoin, Litecoin

Origins and Trends of Virtual Currencies

The first progress toward a VC was made by the cryptography researcher David Chaum, who used cryptographically signed tokens.[10] This and later related proposals paid significant attention to an untraceable, anonymous currency issued centrally and backed by banks or other institutions (who would enjoy a certain amount of trust by users). Digicash, the company Chaum started, managed only a three-year trial at a single bank, which subsequently was not pursued.[11]

Early Systems

VCs have been in use well before the invention of Bitcoin, though they were not decentralized. Digital gold currency and similar systems comprised the first wave of VCs that were created and used. Began in 1996, e-gold was a precursor to the type of system proposed by Chaum; it used a central account structure to track and transfer certificates backed by gold in a central repository with no guaranteed cryptographic security and anonymity, mainly as a function of trust in those running the e-gold system.[12] Since e-gold was outside the financial regulatory system, it offered effective anonymity and security, guaranteed by trust in the operating practices of the companies running these systems. Similar systems, such as Liberty Reserve, WebMoney, and Perfect Money, were frequent targets for illegal activities, both by users abusing the anonymity and the relative ease of transfer beyond the control of regulators and by the operators, who ran Ponzi schemes and

---

[10] See David Chaum, "Blind Signatures for Untraceable Payments," in David Chaum, Ronald L. Rivest, and Alan T. Sherman, eds., *Advances in Cryptology: Proceedings of Crypto 82*, Berlin: Springer-Verlag, 1983, pp. 199–203.

[11] See David Chaum, Amos Fiat, and Moni Naor, "Untraceable Electronic Cash," in Shafi Goldwasser, ed., *Advances in Cryptology—CRYPTO '88: Proceedings*, Berlin: Springer-Verlag, 1990, pp. 319–327, and Julie Pitta, "Requiem for a Bright Idea," *Forbes* online, November 1, 1999.

[12] Doug Jackson, the founder of e-gold, said "practically speaking, e-gold was the opposite of anonymous," as quoted in Kevin Dowd, "Contemporary Private Monetary Systems," self-published paper, August 2013. There was cryptography used in connections to the system, but it was not an intrinsic feature of the currency.

other fraudulent enterprises.[13] All of these systems are distinguished by their centralized authority structure: in order to support illicit activities, actors would have to trust the currency proprietors to maintain their anonymity and security (which some of these currencies historically have done).

Bitcoin

The primary interest regarding VCs in the national security–policy community has been on Bitcoin, in particular with respect to Bitcoin's wide use and its perceived security and anonymity. Bitcoin was introduced in 2009 and exists outside the control of a single company or government; it is defined and controlled by a decentralized group of users executing the Bitcoin protocol over the Internet as described below.[14]

As of June 2015, there were about 14.2 million bitcoins in circulation, with a total market capitalization of $3.5 billion (at an exchange rate of about $240 per bitcoin); this is down from the highest market capitalization of nearly $14 billion in March of 2013 (at a rate of $1,150 per bitcoin). There are currently over 110,000 Bitcoin transactions per day, with a roughly linear increase in transactions from June 2012, at which point there were around 20,000 transactions per day.[15]

The central technological feature of Bitcoin is a global public ledger containing all Bitcoin transactions ever made. The ledger itself

---

[13] See Dowd, 2013.

[14] See Nakamoto, 2008. Bitcoin refers to both a new type of algorithm for a secured public ledger, called the *block chain*, and to the tokens, called *bitcoins*, that are tracked by the ledger and are used as currency. For an excellent, in-depth review of Bitcoin, related VCs, and the academic literature examining them, see Joseph Bonneau, Andrew Miller, Jeremy Clark, Arvind Narayanan, Joshua A. Kroll, and Edward W. Felten, "Research Perspectives on Bitcoin and Second-Generation Cryptocurrencies," *Proceedings of IEEE Security and Privacy 2015*, San Jose, Calif.: IEEE Computer Society, May 2015. For a more policy-oriented introduction, see Edward V. Murphy, M. Maureen Murphy, and Michael V. Seitzinger, *Bitcoin: Questions, Answers, and Analysis of Legal Issues*, Washington, D.C.: Congressional Research Service, August 14, 2015.

[15] See Blockchain, "Market Capitalization," undated (b). Note that the data presented here may fluctuate wildly. In addition, it is difficult to estimate the percentage of "legitimate" transactions versus those executed for criminal purposes.

comprises a sequence of so-called *blocks*; each block contains a list of transactions, as well as the *hash*, or digital signature, of the previous block created (hence the term *block chain*) for the ledger, since each block is chained to the previous one. The block chain is distributed to all computers running the Bitcoin protocol; therefore all nodes in the Bitcoin network have a copy of all transactions ever made. Participants jointly validate new transactions, block by block; more technically, this process is a *decentralized consensus protocol*, where the consensus is whether or not to include the new block into the block chain.

Identities in the Bitcoin protocol are cryptographically generated addresses. Loosely, each transaction is a transfer order from one address to another.[16] The ledger is the recorded history of each of these transactions. A new transaction is allowed to process if the ledger reveals that the sender address had enough of a balance to transfer the proposed amount to the receiver address. By submitting the transaction and having it included in the block chain and acting as the ledger, the corresponding new balance is publicly included in the ledger and reviewable for all future transactions. Accordingly, the number of bitcoins a user owns is the total number of bitcoins associated to the address(es) that the user has access to, and the reason that Bitcoin is said to be "anonymous" is that identity-privacy of bitcoin ownership is maintained through the inability to link an address to an identified user.

It is worth stressing that a user does not *own* bitcoins. Rather, a user has the right to spend the number of bitcoins that are associated with the various addresses they are able to access. Accordingly, a *wallet* of bitcoins is actually the requisite information proving ownership of a Bitcoin address, which in turn allows that user to spend bitcoins associated with that address. Specifically, these addresses are based on a public/private key pair generated cryptographically. The private key allows the coins to be spent in a new transaction. It is conceptually similar to having an address with a locked mailbox; anyone can deliver mail, but only someone with the key can take letters out and send them to a new address, thereby transferring or spending them. In this case,

---

[16] More generally, a single transaction can be conducted from at least one address to at least one other address.

no one necessarily knows who has the key, and the mailboxes are on the block chain.

The anonymity of Bitcoin users, or lack thereof, is a critical component of the currency; see the discussion of VC anonymity in Chapter Four for a detailed discussion of this issue, with a particular focus on Bitcoin.

The blocks of records of correct transactions are validated by employing significant computing power through a process called *mining*; those performing the computations are called *miners*.[17] Mining is successfully completed for a block when a miner has successfully found the correct input to a complex mathematical function called a *hash function*, which effectively binds the validated block to the block-chain transactions. An important feature of the Bitcoin infrastructure is that it is extremely computationally difficult to alter newly validated blocks once bound to the chain, thus preventing changes to the transaction history. For a miner to find a correct input to the hash function, he or she must effectively guess the input at random. This is because finding the input any other way is computationally infeasible (due to the security guarantees of the hash function). In practice, these guesses are made through harnessing many thousands of computing processors. A correct guess is published, providing so-called *proof of work*, because it proves that a miner worked hard to find the input (since finding the input required significant computational work). Other users can easily validate that the miner has found the correct input to validate the block; once verified, the miner is rewarded with bitcoins (in practice, this reward transaction is included in the block that the miner validates).[18] Accordingly, the only way to acquire new bitcoins

---

[17] The validation process is done by checking if the hash of the transactions, plus an additional "nonce" value, conforms to a specific format. The hash function is computationally expensive to run, and any specific set of transactions plus a specific nonce has a very low probability of matching the format. Because of this, miners try many different nonce values, hoping to find one that will validate the block.

[18] In reality, a miner who successfully validates a Bitcoin block obtains bitcoins both from the mining reward process as described above as well as from so-called transaction fees that can be included in every Bitcoin transaction. The number of bitcoins obtained via mining rewards is designed to decrease over time, reaching zero around 2140; the theory is that

is to either mine it or conduct a transaction with another user who already has bitcoins, such as with an online-exchange service to transfer government-backed currency for Bitcoin.[19]

Bitcoin's decentralized, mining-based infrastructure requires that many users dedicate significant resources in order to maintain and secure the overall system. The ability of users to transact in bitcoins depends on the decentralized system's ability to consistently and securely add new blocks to the block chain, thereby validating individual transactions. At the same time, the mining process has become increasingly computationally intensive as the computational difficulty of mining bitcoins was designed to increase with miners. Today, to have a meaningful chance of successfully mining, special-purpose hardware that is specifically optimized for Bitcoin mining is needed.[20]

Chapter Four contains additional discussions about the Bitcoin system, including an examination of the security and anonymity of Bitcoin as well as a discussion of how Bitcoin, and related VCs, can be used for common transactions on devices such as smartphones.

Virtual Currencies After Bitcoin: Altcoins

Bitcoin is not the only VC that a non-state actor might choose to use or build upon for their own VC deployment; many other currencies have built upon the foundational ideas of Bitcoin that a non-state actor might also leverage.

Following the release of Bitcoin, and its subsequent wide adoption and interest, many new projects were launched, a selection of which are represented in Table 1.1. These were based on either the architecture or,

---

transaction fees will correspondingly increase to maintain the economic incentivization of mining, which secures the entire Bitcoin system.

[19] It should be noted that this is a very high-level description of Bitcoin. An interested reader should consult other sources for a more detailed description. See, for instance, Bitcoin Help, homepage, undated; see also Bitcoin Wiki, homepage, August 13, 2015b.

[20] For further discussion, see Michael Bedford Taylor, "Bitcoin and the Age of Bespoke Silicon," paper presented at the *International Conference on Compilers, Architecture, and Synthesis for Embedded Systems (CASES)*, Montreal, Quebec, September 29–October 4, 2013.

Table 1.1
Examples of Appcoins and Block Chain Applications

Examples	Introduced	Application
NameCoin[a]	April 2011	DNS-like storage in block chain
Mastercoin[b]	January 2012	Planned market, smart contracts
Nxtcoin+[c]	November 2013	Asset exchange
Ripple[d]	December 2012	Inter-bank transactions
MaidSafeCoin[e]	April 2014	Anonymous, secured cloud computing (non–block chain)

[a] Namecoin, homepage, undated.

[b] See J. R. Willett, *The Second Bitcoin White Paper*, vs. 0.5 (Draft for Public Comment), self-published paper, undated. Also, see GitHub, "Omni Protocol Specification (formerly Mastercoin)," undated.

[c] See Nxt Wiki, "Whitepaper:NXT," modified July 13, 2014.

[d] See Ripple, "FAQ," undated.

[e] The network is still in pre-beta public testing as of February 2015.

in most cases, a near-total replication of the source code from Bitcoin. Because the block chain is specific to the Bitcoin network, these "altcoins" used new block chains, with various modifications to the protocol. Many of these were effectively Ponzi schemes, with the creators using them to pump-and-dump the new currency, or in other ways that were never intended as legitimate currencies.[21]

We highlight three classes of noteworthy alternatives to Bitcoin as a currency; the first, Pure Altcoins, primarily modified the financial and cryptographic details of Bitcoin. This included currencies that minted coins more rapidly or used different hash functions to vali-

---

[21] Coins have been launched as jokes (e.g., Dogecoin, Pizzacoin, Beercoin) or as proofs of concept and learning exercises (e.g., GeistGeld, Tenebrix). In one case—Liquidcoin—it was announced explicitly as "speculation based" (see Bitcoin Forum, "[RELEASE] Liquidcoin (Speculation Based)," discussion thread began January 18, 2012). In the case of Dogecoin, the joke currency has become less of a joke, with a current market capitalization of $13,874,871 as of February 24, 2015 (see Bitcoin Wiki, "Comparison of Cryptocurrencies," December 24, 2014, and CoinMarketCap, "Crypto-Currency Market Capitalizations," September 30, 2015a).

date the block chain.[22] Yet other new coins altered the method of validating more drastically, replacing proof of work with other schemes.[23] Prominent altcoins include Litecoin,[24] which has a faster hashing process than Bitcoin; Dogecoin, which started as a humorous creation not meant to be taken seriously, then became gradually more accepted; and Peercoin, which uses a hybrid approach to mining that uses an alternative to Bitcoin's proof-of-work system.[25]

The second category, which we will call Anonymous Coins, used additional new cryptographic techniques or protocol to create greater anonymity than Bitcoin offers. This has either been in the form of new altcoins that allow for or enforce a level of anonymity in the protocol or various Bitcoin add-ons using a technique called CoinJoin; see Chapter Four's discussion on VC anonymity for more information about Anonymous Coins.

Most recently, the majority of new effort has been focused on a third category, so-called Appcoins, which use block chains for other purposes. While many Appcoins can be used as currencies and are useful for various types of financial transactions, they create and rely on a more complex infrastructure and do not differ greatly from other VCs in the aspects most relevant to this this report.[26] This new category is interesting because it points to new technological applications of the block chain, though it may be a misnomer to term this category

---

[22] A variety of hash functions and combinations of hash functions have been proposed, largely revolving around concern about centralization of mining power due to application-specific integrated circuit (ASIC)–based mining. Similarly, alternative schemes, such as proof-of-stake, or computing Cunningham chains in Primecoin, have been created. All of these have important pros and cons, but the details are not relevant for most of the following discussion.

[23] For a list of these currencies, see Altcoins, homepage, undated; see also Bitcoin Wiki, "Comparison of Cryptocurrencies," December 24, 2014.

[24] Litecoin, hompage, undated.

[25] Peercoin also uses a so-called proof-of-stake mining system; see Sunny King and Scott Nadal, "PPCoin: Peer-to-Peer Crypto-Currency with Proof-of-Stake," self-published paper, August 19, 2012.

[26] See Chapter Five for further discussion about the implications of VC technology.

as a *currency* due to its intended purposes. (See Chapter Five for further discussion on possible future applications beyond VCs.)

Having given an overview of the VCs and some of their design choices, we now highlight the implications of a particularly important design choice: how to structure the authority mechanism from the centralized structure of older VCs (such as WebMoney) to the fully decentralized structure of Bitcoin.

Authority (De)centralization and Implications for Virtual Currency Design

Perhaps the most prominent design choice in a VC is how centralized its authority mechanism should be. The earliest VC designs, such as Chaum's, had centralized authority mechanisms: there is a central server that ensures that security properties, such as double spending and counterfeiting, do not occur. Drawbacks of such architectures are that they require at least some trust in the central authority (for example, that they do not simply ignore incoming transactions) and that they can be vulnerable to a single point of failure or present a single target for attack. For instance, the M-PESA system, a currency-transfer mechanism that relies only on text messages to conduct transfers in countries such as Kenya, is centralized at the cellular provider; all it would take to disrupt M-PESA is to degrade the cellular network of a particular country (or selected servers of the provider). It is worth noting that non-state actors such as the Islamic State of Iraq and the Levant (ISIL) are unlikely to care about how centralized a currency is from a fiscal policy perspective; however, vulnerability to cyber attack could be a significant concern.

Bitcoin and the vast majority of the second-generation VCs have decentralized authority mechanisms. There is no central server or service, and any user can and do contribute resources to the authority-mechanism process. Such decentralized structures inherently require more public information about users and transactions because each participating user in the authority mechanism must be able to have enough information to contribute meaningfully. In addition, consensus may take time because many users must agree on the best course of action (otherwise small groups of malicious users can break the security of the decentralized scheme). On the other hand, even if some users contributing to the decentralized authority are malicious, they still cannot impede correct behavior on the part of the overall decentralized system due to its consensus-verification system. It is this resilience, and lack of required trust, that has attracted many users to Bitcoin and other decentralized VCs.

There is a middle ground between the two alternatives: so-called semi-centralized VCs, where the authority mechanism is distributed among a restricted set of participants (e.g., ten total) and only when a sufficiently large fraction of them collude would any private information be revealed or would security be violated. This approach may be useful where there are a small number of high-security users who are trusted not to collude with one another; one example might be the central banks (or military units) of multiple countries that may not have completely trusting relationships with one another. The benefit of semi-centralized VCs is that they balance the trust and single-point-of-failure issues with the centralized model and the mass-dispersal issues with the decentralized model. To date, the existence of semi-centralized VCs is largely theoretical;[27] only Ripple may be said to have a fully semi-centralized authority mechanism, and Ripple is not designed to protect user privacy in a meaningful way (for more details, see the discussion on VC anonymity in Chapter Four).[28]

Having discussed the current state of VCs, we will now investigate the extent to which non-state actors are currently using VCs as well as a brief examination of previous politically motivated VC deployments.

---

[27] See, for instance, Karim El Defrawy and and Joshua Lampkins, "Founding Digital Currency on Secure Computation," *CCS '14: Proceedings of ACM SIGSAC Conference on Computer and Communications Security*, March 2014, pp. 1–14.

[28] The VC Dash (formerly Darkcoin) has a hybrid structure where anonymity is guaranteed by a semi-centralized architecture, but most other elements of the currency are governed by a decentralized architecture; see Dash, homepage, undated (a), and Dash, "Masternodes and Proof of Service," undated (b).

Virtual Currencies and Non-State Actors

In this section, we give a brief overview of non-state actors' use of VCs, particularly for criminal purposes, as well as examine previous instances of politically motivated VC deployments.

There is ample evidence that organized non-state actors—especially cybercriminals—use existing VCs.[29] There does not seem, however, to be significant evidence that these actors are regularly conducting standard economic commerce using a VC; rather, VCs are only used as a means of secure, anonymous currency transfer for specialized services. That is, there is no evidence that organized (i.e., nefarious) groups have developed and *deployed* VCs, but there is evidence that some have *exploited* currencies such as Bitcoin for illegitimate transactions.

One of the more common criminal uses of VCs, particularly Bitcoin, is for ransom ware, where cybercriminals encrypt a victim's data and only release it upon payment in a VC, generally Bitcoin.[30] Another common usage is for the purchase of illicit goods (e.g., drugs) on online services similar to Silk Road.[31] This differs from a VC used for everyday commerce, which requires a markedly different physical payment infrastructure that would enable payments at actual physical vendors rather than just websites; technology to enable such payments include smartphones (see Chapter Four's discussion on VC deployability).

There is little evidence that terrorists are using VCs on a meaningful scale, particularly as compared with criminal organizations. The two most-cited examples are two postings by (purported) ISIL sup-

---

[29] See, for instance, Samani, 2013a and 2013b.

[30] See, for instance, Federal Bureau of Investigation, "Ransomware on the Rise, FBI and Partners Working to Combat This Cyber Threat," January 20, 2015.

[31] For one analysis of the Silk Road, see Nicolas Christin, "Traveling the Silk Road: A Measurement Analysis of a Large Anonymous Online Marketplace," *Proceedings of the 22nd International Conference on World Wide Web (WWW 2013)*, Rio de Janeiro: World Wide Web Conference, 2013, pp. 213–223.

porters urging fundraising via Bitcoin.[32] Aaron Brantly of West Point has noted:

> There is sufficient evidence to suggest that terrorists are considering and, in limited instances, using digital currencies such as Bitcoin to finance activities. While these tools have gained in popularity, in recent years their expansion into various terrorist organizations has been slow and deliberate and has not matched pace with transnational criminal uses of these same technologies.[33]

This situation may well change in the future, however, if non-state actors feel they have more to gain—politically, economically, or operationally—by moving toward increased VC usage.

Recently, there have been cases of politically motivated VC deployments to replace existing sovereign physical currency in a sovereign country (with or without government approval). Auroracoin was deployed in Iceland by an unknown source in March 2014 as a means to provide a currency that would be less susceptible to inflation and not subject to government regulation.[34] Derek Nisbet introduced Scotcoin as a new independent Scottish currency.[35] Ecuador is examining the potential of using a VC as an alternative to physical currency.[36] It should be noted that, in the Iceland and Scotland example, the legitimate government did not explicitly sanction the VC deployment, while

---

[32] See Taqi'ul-Deen al-Munthir, "Bitcoin wa Sadaqat al-Jihad: Bitcoin and the Charity of Violent Physical Struggle," self-published article, August 2014, and Adam Taylor, "The Islamic State (or Someone Pretending to Be It) Is Trying to Raise Funds Using Bitcoin," *Washington Post* online, June 9, 2015.

[33] See Brantly, 2014, p. 1.

[34] See Auroracoin, "Why Iceland? Many Governments Have Abused Their National Currencies, but Why Is Iceland Such a Good Place for the First National Cryptocurrency?" undated.

[35] See Folding Coin, "Announcing Scotcoin," February 5, 2015; Alex Hern, "Bitcoin Goes National with Scotcoin and Auroracoin," *Guardian* website, March 25, 2014; and Giulio-Prisco, "An Independent Scotland Powered by Bitcoin?" *CryptoCoinNews.com*, September 17, 2014.

[36] See Nathan Gill, "Ecuador Turning to Virtual Currency After Oil Loans," *Bloomberg News* online, August 11, 2014.

in Ecuador, the government seems to have supported the effort. To date, no replacement-VC deployment has enjoyed widespread adoption.

One of the main purposes of this report is to examine the key challenges that, if overcome, would enable non-state actors, including terrorist groups, to leverage VCs for their political, economic, and/or operational gain. While a non-state actor might choose a more standard paper currency over a VC, changes in perception to VCs in the future, particularly in terms of trusting VCs as a secure, resilient, and available currency, may greatly increase the likelihood of adoption. In particular, support by an allied nation-state with cyber sophistication may greatly influence a non-state actor toward VC deployment.

CHAPTER THREE

Can Virtual Currencies Increase Political Power?

This chapter examines the potential for non-state actors to use VCs to increase their political and/or economic power by virtue of deploying a VC to use as a currency for regular financial transactions. Based on our analyses of the social and political underpinnings of non-state actors' use of currencies, controlling their own currencies can provide non-state actors, such as insurgent groups, with an important tool for increasing their political and economic leverage in contested territories.

Historically, insurgents have issued new currencies in an effort to assert their political and economic control. ISIL's declaration on November 13, 2014, that it will issue its own commodity-based currency fits within this trope.[1] ISIL's choice of a commodity-based currency rather than a VC may be a result of the difficulties involved in deploying a VC in a politically contested territory characterized by relatively low physical infrastructure and low penetration of communications-technology platforms such as smartphones. As discussed in the previous chapter, ISIL's stated intention to use a gold- and silver-based commodity currency also emphasizes the economic credibility conveyed by a currency whose value can be established on international commodity exchanges.

To date, VCs have not been used successfully on a large scale as a full competitor to countries' fiat currencies. Unsurprisingly, given their large technological-infrastructure requirements, VCs have not been the medium of choice for insurgents involved in civil conflicts.

---

[1] See Borzo Daragahi, "ISIS Declares Its Own Currency," *Financial Times* online, November 13, 2014.

Some separatist-movement supporters in developed countries, such as Scotland,[2] have issued VCs (e.g., Scotcoin), but without popular support. Auroracoin was launched in Iceland as a means to contest the government's strict capital-controls regime. As such, it was not undertaken as a vehicle for insurgency or separatism, but did constitute a political protest against the government's macroeconomic policies. The developers adopted the tagline, "a nation breaks the shackles of a fiat currency."[3] As a VC, Auroracoin represented an interesting experiment that nevertheless failed to attract users in that the Icelandic population appeared unwilling to switch from króna to Auroracoin, despite Iceland's capital-controls regime.

These examples demonstrate both the technological feasibility of a non-state actor deploying a VC, as well as the challenge faced by a non-state actor to encourage societal participation in a new VC when traditional currency options remain available. We expect non-state actors will be most likely to get people to use a new VC when the non-state actor has sufficient territorial control and governance capacity to enforce the use of its VC.

Non-State Currencies Emerge When State Currencies Do Not Meet Groups' Needs

With the wealth of attention enjoyed by VCs such as Bitcoin, one might think that VCs were playing a significant role as a new medium of exchange for day-to-day transactions in countries such as the United States. A recent Bloomberg article touted the growing popularity of Bitcoin, reporting:[4]

> Consumers are embracing the digital currency . . . Parents are dispensing allowances in Bitcoin to teach their kids to be digi-

---

[2]   See Scotcoin, homepage, undated.

[3]   See Auroracoin, undated.

[4]   See Olga Kharif, "Bitcoin: Not Just for Libertarians and Anarchists Anymore," *BloombergBusiness.com*, October 9, 2014.

tal citizens. Marijuana smokers are buying buds from Bitcoin-enabled vending machines. Consumers in emerging markets such as Brazil and Russia are starting to use Bitcoin to hedge their volatile currencies.

The overall demand for VCs as fully fledged competitors to centrally managed fiat currencies in countries with strong state capacity and stable macroeconomic policies, however, is relatively small. Central banks and governments in developed countries have assessed the monetary-control risk posed by VCs circulating in their areas of responsibility to be low, at least at current and foreseeable levels of VC circulation.[5]

There are two conditions under which VCs are likely to gain traction as market actors' preferred currency option. The first condition is that the central authority does not provide a stable macroeconomic environment and, as a result, the territorial fiat currency is nonexistent or its value becomes unstable.[6] The European Banking Authority highlights this environment as one the key findings of their report on VCs.[7]

> In jurisdictions where financial services are not widely available, where users have a high risk profile, where the national currency is not convertible into other [fiat currencies], where financial services are too expensive for individuals, or where the administrative burden for obtaining an account is high, VC schemes provide an alternative way for individuals to achieve the same end: accessing commerce and effecting payment transactions.

---

[5] See European Central Bank, 2012, and Murphy, Murphy, and Seitzinger, 2015.

[6] It is important to note that a VC is not the only monetary alternative to a fiat currency in a territory lacking a stable macroeconomic environment. Market participants could also engage in barter, develop their own scrip-based community currency, or use another country's currency. The U.S. dollar has been used extensively outside of the United States. This report focuses on the feasibility of non-state actors deploying a VC. It does not provide a full assessment of the tradeoffs across the full menu of alternative currency arrangement options available to non-state actors.

[7] See European Banking Authority, 2014.

There are many reasons why territories may lack a stable national currency. The territory may be part of a failed state with no functioning government; it may be part of a country in the midst of a civil conflict; or it may be a country with a stable government, but unstable macroeconomic policies or policies that freeze out economic participation by a large fraction of its population (e.g., countries with a large black market). In an environment in which the central authority cannot safeguard the stability and accessibility of a fiat currency, a non-state actor–sponsored VC may provide a viable solution.

The second condition in which VCs may play an important role in is building and maintaining communities. At the local level, many communities have set up regional exchange trading systems.[8] Local currencies broaden a community's exchange infrastructure beyond economic exchanges to also support social, ethical, and environmental dimensions valued by the community. Most community currencies are geographically constrained and circulate side by side with national currencies.[9] Their value in a particular community is specific to the goals of that community, and their creation may reflect payment for services rendered solely for that community. Examples of community currencies are Ithaca Hours and Salt Spring Dollars,[10] or more global currencies, such as frequent-flier miles. While most communities use paper currencies, there have been a few forays into community VCs. The Totnes Pounds system supports both a paper currency and electronic accounts.[11]

---

[8] Although we focus here on recent examples of community currencies, Christine Desan examines the importance of community exchange systems and the role of community stakeholders in medieval English communities (Christine Desan, Making Money: Coin, Currency, and the Coming of Capitalism, Oxford: Oxford University Press, 2014).

[9] See Jerome Blanc, "Thirty Years of Community and Complementary Currencies," *International Journal of Community Currency Research*, Vol. 16, 2012, pp. D1–4.

[10] See Ithaca Hours, homepage, undated, and Salt Spring Dollars, homepage, undated.

[11] See Totnes Pound, homepage, undated.

David Vandervort and colleagues at PARC identify Mazacoin,[12] the purported national currency of the Lakota nation, and Irish Coin,[13] a community coin developed to promote the Irish tourism industry, as two exemplars of community VCs.[14] As technology for VCs improve, community VCs may become more common.

Most local communities that have adopted community currencies have done so within the structure of a well-developed financial system, i.e., in a stable, generally democratic country or international system with stable macroeconomic policies[15] or shared norms of behavior. Not all non-state actors, however, choose to develop alternative currencies that function complementarily with their country's fiat currency. Many non-state actors, such as separatist and insurgent groups, as well as contested regions issue their own currencies to highlight their economic sovereignty and to solidify their economic control in territories under their jurisdiction or lands they wish to control.[16] For example, the autonomous region of Somaliland has its own shilling, and the autonomous region of Transnistria has its own ruble. A whites-only town in South Africa called Orania uses a currency called the ora.[17] In addition to ISIL's declaration to launch its own currency, the aspirational central bank of Barotseland in 2012 declared the introduction of the Barotseland mupu in 2012.[18]

---

[12] See Mazacoin, homepage, undated.

[13] See Irish Coin, homepage, undated.

[14] See David Vandervort, Dale Gaucas, and Robert St. Jacques, "Issues in Designing a Bitcoin-Like Community Currency," paper presented at the Second Workshop on Bitcoin Research, San Juan, Puerto Rico, January 30, 2015.

[15] Damjan Pfajfar, Giovanni Sgro, and Wolf Wagner, "Are Alternative Currencies a Substitute or a Complement to Fiat Money? Evidence from Cross-Country Data," *International Journal of Community Currency Research*, Vol. 16, 2012, pp. 45–56, found that the use of community currencies is positively associated with stability in the country's fiat currency, financial sector development, and overall economic development.

[16] See Daniel Treisman, "Russia's 'Ethical Revival': The Separatist Activism of Regional Leaders in a Postcommunist Order," *World Politics*, Vol. 49, No. 2, 1997, pp. 212–249.

[17] See Wikipedia, "Ora (Currency)," April 27, 2015.

[18] See Barotseland Free State, *Barotseland Mupu Currency Act of 2012*, February 28, 2012. Barotseland is a contested territory between Zambia and Angola. The central bank has

Using the example of an insurgent group with contested territorial control over a region, insurgent groups have three options when adopting a currency.

Their first option is to adopt a commodity-based currency in which the currency in circulation is the commodity itself (e.g., gold coins). This is ISIL's stated strategy. The key benefit of this option is that the credibility of the currency is backed up by the intrinsic value of and international market for the commodity. There is no need to trust the monetary authority that gold or silver will retain its value. A key limitation of this option is that it is difficult for most insurgent groups to amass sufficient supplies of gold and silver to implement this type of currency.

Their second option is to adopt another country's currency. This option can range from circulating the pre-existing currency directly in the local economy (e.g., dollarization) to minting a new currency that is backed 1:1 with reserves of another country's currency (e.g., use a currency board). Leaders of the self-proclaimed Donetsk People's Republic have attempted to set up a ruble zone in eastern Ukraine. The costs and benefits of this option are somewhat similar to those of a commodity-based currency. The new currency gains credibility based on the stability of the issuing country's currency and the adoption of the currency throughout the territory. If the value of the issuing country's currency falls, then so will that of the new currency. The feasibility of this option depends on the insurgent group's ability to amass sufficient supplies of the adopted currency to issue it in their territory or use it to back up their own currency. This is a much easier hurdle to overcome, especially with the support of the issuing country; indeed, support by another country, particularly when that country possesses cyber sophistication, is one key enabler for non-state actor deployment (see Chapter Four for further discussion).

Their third option is to adopt its own currency. In this option, the currency may not necessarily be backed 1:1 by a commodity or by stocks of a reserve currency. One example of this occurred when

---

released samples of mupu bills, but does not have the resources to issue and maintain a stable currency.

separatist authorities in Somaliland introduced the fiat, paper-based Somaliland shilling without being explicitly tied to a commodity or reserve currency. Accordingly, the benefit of this option is that the insurgent group may require smaller reserves of commodities or foreign exchange to roll out their new currency. The drawback of this option is that there is no intrinsic value built into the currency at its outset.[19] A fixed-exchange rate peg may help to combat volatility in the currency's value; however, unless the market believes the currency is accurately valued or the group has the foreign exchange reserves to defend their exchange rate peg, the group may be unable to maintain the currency's value.[20]

Non-State Currencies Are Not Likely to Be VCs Now, But Could Be VCs in the Future

Although some separatist groups have attempted to adopt their own currencies, we do not expect VCs to be their preferred format in the near term. There are three main reasons why deploying a VC may pose greater difficulties than a paper- or commodity-based option. The first reason is that most insurgent organizations currently lack the skills necessary to deploy a VC. Most insurgencies occur in politically contested territories characterized by low physical infrastructure and low penetration of communications-technology platforms, such as smartphones. Although a prominent debate over the use of VCs as a source of development capital in areas of low economic development has sparked interest in the development community, the need for advanced cellular phones (i.e., smartphones) for VCs has impeded implementation. In contrast, M-PESA, a mobile phone money-transfer system (and a convenience mechanism rather than a VC), works well in Nigeria, as

---

[19] It should be noted, however, that hesitation with respect to new currencies can be overcome through careful and proper rollout of a new currency. The most famous example is Brazil's rollout of the fiat currency réal (R$) in 1994, which was carefully managed to replace the old currency to defeat hyperinflation.

[20] Alternatively, the group may declare the currency is not convertible, in which case the fixed exchange rate statement may simply serve a symbolic purpose.

it has the active support of the Nigerian government and much lower technological requirements. We discuss the technological requirements and challenges for a VC in greater detail in the next chapter.

The second reason is that the monetary rules underlying a VC need to be specified and maintained. These rules specify such currency characteristics as how actors will be incentivized to create and secure the currency, whether the money supply will be capped or continue to grow, and whether the money will be geographically constrained or can be used globally. In a centralized system, these rules are set and enforced by the central authority. In a decentralized system, some form of rule-adoption process is needed. Community currency systems have often failed as the communities included in the currency system grew too large to adjudicate the rules-adoption-and-enforcement process effectively.[21] VCs such as Bitcoin tend to be decentralized systems in which rules governing the currency and incentivizing its expansion and maintenance are specified by their designers but are subject to consensus decisionmaking at the protocol level, between the servers. Decentralization is often a key characteristic of a VC's resilience. Maintaining control over the rules of the game, however, is a source of vulnerability for a decentralized VC with relatively small circulation. An insurgent group that sets up a VC would face a tradeoff between a centralized authority structure that would not be vulnerable to rules changes triggered by a majority of currency holders, but might be more vulnerable to external (and internal) attack. By contrast, a decentralized authority structure may be more resilient to external attacks, but less amenable to rules changes.

The third reason is that, at least at the outset, users' trust in new currencies tends to be low.[22] Users need time to become familiar with and feel assured by the system and the stability and the ease of use of the currency. We expect this will be exacerbated for new VCs.

---

[21] See Georgina Gomez, "Sustainability of the Argentine Complementary Currency Systems: Four Governance Systems," *International Journal of Community Currency Research*, Vol. 16, 2012, pp. D80–89.

[22] See Matthias Kaelberer, "Trust in the Euro: Exploring the Governance of a Supra-National Currency," *European Societies*, Vol. 9, No. 4, 2007, pp. 623–642.

Implementing a new currency of any type is difficult. It entails large technological, economic, and logistical challenges. Particularly for insurgent groups that choose to deploy their own fiat currency, the trustworthiness of the currency is an important component of its success. A low initial penetration of VCs in day-to-day economic life will increase users' suspicion of currencies deployed through this technology. Although paper currency may require greater physical infrastructure and be less resilient to physical attack, in the near term, paper currencies will be far more acceptable and inherently trustworthy for the population than VCs. That said, populations' suspicions of VCs will erode as they become more familiar with them. In a territory in which a VC is the only medium of exchange, economic necessity may force people to accept VCs where they would have otherwise rejected them. That is, everything else being equal, an insurgent group is more likely to choose a paper currency (whether or not it is backed by government-controlled commodities) over a VC *today* in order to increase the populations' trust in the currency, but there could be a shift in attitude as the technologies that underlie VCs become more prevalent and trusted.

CHAPTER FOUR

Technical Challenges to Virtual Currency Deployment

In this chapter, we examine the technical challenges that a non-state actor might face when deploying a VC. These challenges could potentially be leveraged by opponents, such as the United States, to impede the success of the non-state actor's VC deployment. Some of these technical challenges relate to ensuring that a VC deployment is widespread and usable enough for everyday financial transactions (e.g., buying a soda at the corner store), while other challenges relate to securing a VC deployment so that it is trusted for everyday use. In addition, any entity deploying a VC needs to ensure resilience of the currency against cyber threats by opponents, including the most advanced threats posed by competitor nation-states.

We emphasize that this section is focused primarily on issues of VC deployment rather than exploitation; however, some of the challenges that we examine here, particularly those relating to anonymity, also apply to VC exploitation.

Specific technical challenges facing any actor attempting to deploy a VC for everyday use include:

- Having access to the technological sophistication necessary to develop, deploy, and maintain a VC as a cyber service. In the context of VCs, the technological sophistication required includes competencies in networking, computation, and cryptographic techniques.

- Ensuring that users of the currency have persistent, assured access to their currency while requiring a sufficiently low level of technological sophistication to enable use for everyday transactions.
- Ensuring levels of transaction anonymity demanded by users while ensuring transaction integrity so that buyers and sellers are assured of proper exchange—all without the need for overly advanced technological expertise.
- Protecting the overall integrity (and availability) of a VC against advanced cyber threats, particularly those nation-states that would oppose the non-state actor's VC deployment.

It is important to note that these challenges are not unique to Bitcoin or other decentralized VCs (see Chapter Two for a discussion of the current state of the art). Indeed, it is not clear that a non-state actor would favor the lack of a central authority. Accordingly, one of the main initial decision points for the creation of a VC is how to structure the *authority infrastructure*, i.e., the network of computers executing algorithms that perform the same aggregate functionality that a centrally postured authority would.

Throughout this chapter, we discuss VC deployment as if the non-state actors are acting essentially on their own. Were they to be backed by a nation-state with moderate to sophisticated cyber capabilities, however, that might well change a non-state actor's decision calculus as to whether (and how) to deploy a VC.[1] By convention, we will call adversaries of the non-state actor deploying a VC their *opponents*.

Developing and Deploying a Virtual Currency

One of the main technological barriers for a non-state actor to deploy a VC would be the expertise and general capability necessary to develop and deploy both the currency and the means to transact with it. In

---

[1]   It is worth highlighting that there is ample evidence of state actors supporting non-state actors *generally*. At issue here would be significant support—indeed, direct and sustained coordination—in the domain of cyberspace operations, which, while feasible, seems different in nature than historical and current examples of state-actor support.

principle, the technical sophistication required to develop and deploy a VC are relatively high, but in practice current technologies exist for general adoption to support such a deployment. Further, the main goal is to identify those key issues now that, once overcome, would greatly impact a non-state actor's ability to deploy a VC.

The key components that would require development are: (1) the currency itself, including numerous important design choices; (2) the means of acquiring, maintaining, and transferring currency as part of financial transactions, including the physical means capable of supporting such transfers such as smartphones; and (3) sufficient back-end services and front-end payment-processing systems to support all of these services in a secure and resilient manner.

Developing Software for a Virtual Currency

The difficulty of development of computer software for a new VC depends on the degree to which a non-state actor wishes to depart from existing VCs and/or their associated software. At one extreme, a non-state actor could simply use Bitcoin or another existing VC outright as their currency, but this raises the question of how a non-state actor would gain politically or economically from such simple adoption. At the other extreme, a non-state actor may decide to create an entirely new currency from scratch; this would require access to software developers with significant skill. A compromise between the two extremes, which is perhaps the most feasible, would be for a non-state actor to create a new VC by using essentially the same software used by an existing VC.

Software developers would have to design software to regulate the currency (e.g., miners for Bitcoin-type decentralized currencies) as well as software applications for everyday users to maintain and transact in the VC; all of this development would have to be usable enough to encourage widespread adoption and use.[2] Given the inherent underlying security required for such applications, one rough (lower-bound) estimate for how sophisticated such a developer would have to be would be on the order of creating custom, widely used encryption soft-

---

2   See Open Hub, "Project Bitcoin Summary," undated.

ware. Indeed, there are very few examples of such software currently in use—and one public case of such software abruptly disappearing from use.[3] Note that if a non-state actor had nation-state backing, including access to that country's cyber experts and developers, such development may be far more feasible. Even in this more ideal situation, however, there are cases of advanced cyber powers having difficultly creating widely deployed cyber services in even the most permissive environments, such as the United States' development of the online exchanges to support the Affordable Care Act. Alternatively, a non-state actor may rely on allied, or paid, "hacktivists," cyber-criminal organizations, or cyber mercenaries.[4] It should also be noted that some non-state actors, in particular terrorist organizations, seem to have at least a limited ability to create secure cyber services, such as encryption platforms.[5]

The most straightforward way of developing a new VC is to repurpose an existing VC—that is, keeping the underlying technological aspects of an existing currency while regenerating it under a new name. We note that this setting is different from using Bitcoin or another VC; the software may be the same, but used as a separate cyber service (whereas above, the non-state actor would *actually use* Bitcoin or some other existing VC). Many existing VCs are repurposed or extensions of Bitcoin (see Chapter Two for more details). In some cases, constructing a new VC requires very few cyber capabilities inasmuch as there exist online services that advertise VC creation services. One possible issue is leveraging old software may have the side effect of importing existing cyber vulnerabilities contained in that software.

---

[3]   See Brian Krebs, "True Goodbye: 'Using Truecrypt Is Not Secure,'" *KrebsonSecurity.com*, May 14, 2014.

[4]   See, for instance, Kaspersky Labs, "The Desert Falcons Targeted Attacks," version 2.0, corporate publication, Moscow: Kaspersky Labs, 2015.

[5]   See, for instance, Recorded Future, "How Al-Qaeda Uses Encryption Post-Snowden (Part 1)," self-published paper, May 8, 2014a, and "How Al-Qaeda Uses Encryption Post-Snowden (Part 2)—New Analysis in Collaboration with ReversingLabs," self-published paper, August 1, 2014b.

Physically Deploying a Virtual Currency

Another significant challenge in deploying a VC is that of physical deployment, i.e., identifying the medium through which the average citizen can transact with their neighborhood vendor. While a computer may be enough for some VC transactions, in order to enable everyday transactions, VC users will need far more portable devices with which to conduct transactions. Unlike paper-currency transactions, the computational complexity of these transactions constitutes a significant barrier to deployment because the average user may not have the existing physical means with which to conduct everyday transactions.

On one hand, the easiest answer to this issue is smartphones, since they already have significant capabilities to both compute and communicate. For instance, Bitcoin has many possible smartphone applications that can be used for transactions.[6] The use of smartphones for VC transactions is hardly novel, indeed, many vendors in developed countries already use their smartphones (or tablets) for standard currency credit-card transactions through applications such as Square.[7]

Depending solely, or primarily, on a smartphone-dominated currency system is challenging for several reasons. The greatest issue is that creating a smartphone-based currency requires that each person who transacts must have a smartphone or equivalent; this is not currently a realistic assumption in any country, let alone developing countries. Another issue is that a currency architecture that relies solely on smartphones, or for that matter any single device, leaves a user extremely vulnerable to currency theft if the device is stolen. For currently conceived VCs, the theft of a password that gives access to the wallet or application allows for the theft of all currency associated with that password. By contrast, for physical currency, a thief is generally limited to cash on hand; automated teller machine (ATM) withdrawal limits; or other limitations, such as personal-check revocation before use.[8] Therefore,

---

6   See, for instance, Bitcoin, "Choose Your Bitcoin Wallet," undated (a).

7   See Square, homepage, undated.

8   The assumption here is that VC transactions are "non-revocable," i.e., once transactions are made, they cannot be undone. In truth, non-revocability is typically only the case for decentralized VCs (and in particular is true for Bitcoin). Revocation can be done technically

any VC that was accessible from a limited number of devices would greatly benefit from advanced security mechanisms, such as biometric verification (which Apple Pay uses)[9] or some other multifactor authentication (such as requiring a Bluetooth link between a phone and an additional required device, such as required by Coin)[10] that would allow for additional security or credential-revocation capabilities.

Using smartphones is not the only means of conducting digital transactions. Indeed, the African (standard) currency-transfer system, M-PESA, has been using non-smart (i.e., "dumb") cell phones for years.[11] It is possible to conduct existing VC transactions with such phones using text messages,[12] but these systems essentially use a central server trusted to maintain a wallet. Due to the high level of trust required in the service provider, it is unclear that adoption of a VC with such a setup would be likely due to trust issues. In principle, one could create a wallet application for non-smartphones, though it is difficult to install such an application in a widespread fashion, since such phones are typically not set up for such remote installations (wallet applications are a security challenge for smartphones in their own right; see the section on cyber threats to VCs later in this chapter).

At the same time, there is evidence of increased mobile-phone usage for financial transactions (not just smartphones), particularly in Africa (see Figure 4.1). Much of this popularity, particularly in

---

if a VC was centralized or semi-centralized. With some regulatory/law enforcement capability built in, theft may not be as crucial an issue, though it would still be highly inconvenient.

[9]   See Apple, "iOS Security, iOS 9.0 and Later," September 2015.

[10]   See Only Coin, homepage, undated.

[11]   See William Jack and Tavneet Suri, "The Economics of M-PESA," second version, self-published paper, August 2010; and Ignacio Mas and Dan Radcliffe, "Mobile Payments Go Viral: M-PESA in Kenya," World Bank website, March 2010. M-PESA is very different from the VCs considered here because it is a means of transfer more than currency; users buy the currency from physical vendors and transfer the currency via text message. The cell-phone provider is trusted to conduct the transactions. In the case of a non-state actor trying to deploy a VC in a denied environment, such an infrastructure would be unlikely to succeed because it presents many points of attack (the single cell service provider, the physical merchants).

[12]   See Blockchain, "Send Via: Send Bitcoins Using Email and SMS," undated (c).

Figure 4.1
Mobile Payment Use

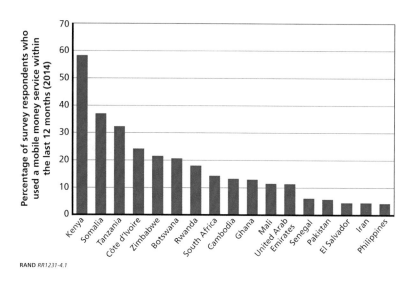

RAND *RR1231-4.1*

SOURCE: World Bank, *Global Findex Database*

Kenya, is due to the adoption of the M-PESA money-transfer scheme, as mentioned above. Figure 4.1 demonstrates that mobile-phone usage for payments is not restricted to Africa: 13 percent of Cambodians, 12 percent of Emiratis, 6 percent of Pakistanis, and 4 percent of Iranians reported using mobile-money services within the last 12 months in 2014. Note that the data in Figure 4.1 is restricted to mobile payments using mobile-money services; such payments are distinct from mobile payments using existing financial institutions (e.g., banks), and therefore should be viewed as a lower bound on mobile-phone usage for financial transactions. As a result, the usage of mobile phones to conduct everyday VC transactions should be viewed as feasible, particularly in the future.

Other means of conducting VC transactions exist beyond mobile phones, but they would require additional hardware, such as USB drives or smart cards.[13] This would require the wide-scale distribution

---

[13] See Bitcoin, undated (a), and Bitcoin Wiki, "Hardware Wallet," August 15, 2015a.

of such devices by a non-state actor, greatly increasing the difficulty of a VC deployment from a purely cyberspace exercise.[14] In addition, such hardware could be subject to a supply-chain attack by a sophisticated opponent (what we will term as *Tier V* or *Tier VI* as outlined in *Task Force Report: Resilient Military Systems and the Advanced Cyber Threat*, published by the Defense Science Board in 2013; see the "Cyber Threats to Virtual Currency" section below as well as Appendix A for further information).

Deployment Challenges for Decentralized Virtual Currencies

Another challenge that would be faced by a non-state actor in deploying a decentralized VC is how to incentivize mining operations, which comprises the key security component of all existing decentralized VCs.[15] If the non-state actor has a poor global reputation (e.g., ISIL), individuals from around the world may be disinclined to mine their currency (e.g., if a hypothetical ISILCoin were a VC). In fact, such mining might conceivably be made illegal in many countries under counterterrorism laws. If the miners were less geographically diverse, this could provide a challenge from a security perspective because such miners might be easier to target; on the other hand, the geographical diversity of diaspora communities may mitigate this issue. Another concern is that if a currency is only transacted in a geographically fixed location (e.g., Iraq and Syria for ISILCoin), potential miners who were not geographically proximate would be deterred from mining because they would have less use for the currency. One solution to this would be to ensure that the VC would be worthwhile for online transactions, but this raises additional challenges, not least of which is that other existing VCs might now have reason to attack the newer VC (for more on this idea, see the section below on cyber threats to VCs).

---

[14] There is some evidence that ISIL has deployed a chip-enabled identification card, though such a card would be less sophisticated than a smart card, which has cryptographic computation capabilities. See, for instance, Arijeta Lajka, "Islamic State Takes a Stab at Legitimacy with Alleged Identification Cards as Forces Lose Ground in Iraq," *Vice News* online, April 16, 2015.

[15] If the currency is decentralized but not mined, alternative means of incentivizing security must be devised.

There are other means to incentive mining. One is to have the mining process perform a public good; this might encourage a broad group of users to mine even if they do not transact in the currency. An example of this is PrimeCoin, where the mining process searches for prime numbers, though it is unclear if users would feel that the public good would outweigh the support given the sponsoring entity.[16] Another technique to incentivize mining is *merge-mining*, which essentially embeds the mining of the new VC within the Bitcoin mining process; the issue with such a process is that Bitcoin could choose to alter their rules to disallow such embedding to occur (since presumably those mining Bitcoin would not view funding terror positively, if that was how the non-state actor was publicly viewed), which would crash the currency.[17]

Virtual Currencies, Adoption, and Value

A major challenge for VC deployment is how to instantiate a VC— namely, going from zero users to an entire community within the geographic area of interest of the non-state actor. Indeed, Bitcoin took four years to gain any significant value as a currency,[18] and it is clear that the vast majority of VC deployments fail to gain acceptance.[19]

Indeed, the best time to target a VC would be during this initial period: trust in the currency may be at a minimum, the skill of cyber defenders may also be at an all-time low, and overall sensitivity to the stability and success of the currency may be very high. As a result, this may be the best time to attempt to disrupt the currency and/or degrade VC user confidence in order to prevent successful deployment.

---

[16] For another example, see Sunny King, "Primecoin: Cryptocurrency with Prime Number Proof-of-Work," self-published paper, July 7, 2013.

[17] See Bonneau et al., 2015, for further discussion. Since the Bitcoin-rules adoption process is largely decentralized, the enforcement of such a prohibited embedding is worth further examination.

[18] See Blockchain, "Market Capitalization," undated (b).

[19] For instance, as of June 2015, only 41 of the 514 listed VCs had market capitalizations greater than $1 million; eight VCs had market capitalizations over $10 million; and three VCs had market capitalizations over $100 million; see CoinMarketCap, 2015 (a).

One potential strategy to mitigate this issue is to back the new currency with a pre-existing VC,[20] a paper-based currency, or a commodity to anchor the VC's value. In particular, if the non-state actor had the support of a nation-state, a VC's value could be set to a fixed exchange rate with the nation-state's currency (or some commodity that the nation-state possesses). This strategy would entail ceding economic control over a VC, thus potentially trading off the speed at which a VC gains acceptance with the political and/or economic benefits of deployment.

Ensuring Anonymity of Currency Use

This section examines the degree at which current VCs afford users sufficient anonymity, how hard it is to de-anonymize some VC transactions, and what level of user sophistication may be required in order to ensure sufficient anonymity. We focus on Bitcoin and its anonymity issues, but also examine how new VCs and associated technology may be able to greatly increase user anonymity.

Investigating VC anonymity is crucial because anonymity is one of the most important properties of any currency, namely that "neither the buyer nor the seller requires knowledge of its history."[21] While

---

[20] See Bonneau et al., 2015, for further discussion, particularly in regard to pegged side chains as well as forking Bitcoin. Such a tie to a previous VC might also solve the mining issue because all mining could be done in the existing VC (e.g., Bitcoin), and then exchanged into the new VC. One example of a currency that is mined in this manner is Zerocash (Eli Ben-Sasson, Alessandro Chiesa, Christina Garman, Matthew Green, Ian Miers, Eran Tromer, and Madars Virza, "Zerocash: Decentralized Anonymous Payments from Bitcoin," paper presented at the 2014 IEEE Symposium on Security and Privacy, San Jose, Calif., May 18–21, 2014a), in that it could rely on an underlying "standard" altcoin and then build on top of it.

[21] See Kenneth Rogoff, "Costs and Benefits to Phasing Out Paper Currency," *NBER Macroeconomics Annual 2014*, Vol. 29, 2015, pp. 445–456. The paper continues:

"There is nothing, however, in standard theories of money that requires transactions to be anonymous from tax- or law-enforcement authorities. And yet there is a significant body of evidence that a large percentage of currency in most countries, generally well over 50%, is used precisely to hide transactions."

much of the focus on VCs is on the anonymity of some illicit transfers (thus making it untraceable by law enforcement or military and/or intelligence organizations), there are other issues that arise from using a VC as an everyday means of transfer. Without sufficient user anonymity, everyday users will be strongly deterred from using any VC for everyday economic transactions due to potential serious privacy breaches. Note that "anonymity" is a broad concept, and attacks to de-anonymize range from highly sophisticated attacks to de-anonymize a single user to hacker attacks in which little effort and sophistication is required to de-anonymize broad groups of individuals.

The following sections first examine VC-anonymity tradeoffs versus VC-authority centrality. We then examine the particular case of Bitcoin, as it is the most prevalent VC and accordingly has been subject to the most examination. Finally, we examine other technologies for Bitcoin as well as newer VCs that may provide sufficient levels of user anonymity.

Anonymity Versus Virtual Currency Centralization

As implied above, assuring everyday user anonymity is a different issue from assuring anonymity from the government by technically sophisticated and determined groups. In practice, however, virtually all VCs do not make this distinction. For decentralized VCs, the two issues are currently inseparable because distinguishing attack source and sophistication is not a design parameter within a decentralized infrastructure. In the centralized VC authority setting, there has been an extensive body of work discussing how to make such a distinction,[22] but in practice, many centralized VCs (e.g., Perfect Money, WebMoney, Liberty Reserve)[23] also protect user identities from, for example, government investigations (though the central VC authorities have to be trusted to do so, whereas cryptographic VCs try to base their security on math-

---

[22] This body of work was initiated by (Chaum, 1983) and continues today. Such schemes typically rely on some notion of trusted party, which decentralized VCs seek to avoid.

[23] See Perfect Money, homepage, undated, and WebMoney Transfer, homepage, undated. See also Brian Krebs, "U.S. Government Seizes LibertyReserve.com," *KrebsonSecurity.com*, May 13, 2013.

ematical proofs). Note that these VCs are exactly the ones targeted by governments because they attempt to provide such anonymity but are ripe for attack due to their centralized architecture (motivating the case for decentralized VCs).[24] Semi-centralized VCs are uncommon, so they are hard to evaluate here. The best-known semi-centralized VC, Ripple, is not designed to be private because it is geared toward financial institutions rather than individuals.[25] At the same time, there has been debate in the cryptography literature that semi-centralized VCs may be the best way forward to maintain security and privacy for individuals while simultaneously allowing for government regulation, but no such VC has been deployed.[26]

"Anonymity": A Bitcoin Case Study

To examine how anonymous a VC may be, we consider the case of Bitcoin; for an introduction to the technical basics of Bitcoin, see Chapter Two. In principle, Bitcoin is pseudonymous because every user is represented by a random, cryptographically generated string of digits, called an *address*, which does not reveal the user's actual identity. If a user does not change his or her address from transaction to transaction, however, then the entire transaction history is completely public to anyone who knows his or her Bitcoin address. This is because the Bitcoin block chain, which is the public ledger, is a public record of every transaction that has ever occurred. Therefore, repeated Bitcoin transactions while reusing the same Bitcoin address poses a serious risk to anonymity. Note that a Bitcoin address can become known by many people in the course of regular transactions by anyone transacting with the user, such as store owners, companies paid, friends who are trans-

---

[24] See, for instance, *United States v. Liberty Reserve*, 13 CRIM368 (S.D.N.Y. 2013), and Department of Justice, U.S. Attorney's Office, Southern District of New York, "Indictment and Supporting Documents: U.S. v. Liberty Reserve et al.," May 28, 2013.

[25] As the Ripple FAQ website says, "Anonymity is not a design goal of Ripple. However, Ripple should provide adequate privacy for most people." See Ripple, undated (a).

[26] See El Defrawy and Lampkins, 2014. The alt-coin Dash (formerly Darkcoin) may be viewed as semi-centralized in some respects due to its Masternode construct, but this is mainly for anonymity purposes—the other functions of the currency are performed analogously to Bitcoin; See also Dash, undated (b).

ferred money, and the like.[27] Put another way, Bitcoin is anonymous in the following sense: It is as though every bank transaction and every bank-account balance is known to anyone with an Internet connection; the only information that is unknown is who owns each bank account, something that can be inferred from user interactions.

Clearly, such "anonymity" is unacceptable for everyday economic life, and therefore additional safeguards must be built in.[28] For many (if not most) existing VCs, including Bitcoin, the current process of maintaining anonymity amounts to learning a degree of cyber-operational security or "tradecraft," which seems unrealistic for the layperson. In the following sections, we examine the means of anonymizing Bitcoin as representative of other VCs for two reasons: first, Bitcoin is the most popular VC with the most efforts to protect information; second, many VCs are built using Bitcoin as their foundation, and therefore many of the efforts to anonymize Bitcoin can be applied for other VCs.

Anonymizing Bitcoin comprises two aspects: (1) anonymizing individual transactions and (2) anonymizing the patterns of transactions. Making individual transactions anonymous is accomplished predominantly by assigning a random pseudonym to each individual. Even with this pseudonym, an individual transaction might be identified by examining the Internet protocol (IP) addresses of the users, thereby revealing the user's entire transaction history. Accordingly, technologies to mask IP could be used if anonymity is desired; the

---

[27] In particular, if an illicit organization were to attempt to fundraise via Bitcoin by publishing their address for others to transfer money to, then the block chain would be a permanent, public record of the pseudonym of every other user who had given money to that organization, unless the donors use some kind of advanced tradecraft; see the following paragraphs for more on this topic.

[28] Indeed, the Bitcoin Foundation (Bitcoin, "Some Things You Need to Know," undated [c]) says the following:

> Bitcoin is not anonymous. Some effort is required to protect your privacy with Bitcoin. All Bitcoin transactions are stored publicly and permanently on the network, which means anyone can see the balance and transactions of any Bitcoin address. However, the identity of the user behind an address remains unknown until information is revealed during a purchase or in other circumstances. This is one reason why Bitcoin addresses should only be used once. Always remember that it is your responsibility to adopt good practices in order to protect your privacy.

Bitcoin foundation recommends the use of such technologies, specifically mentioning Tor.[29] The anonymity of using Bitcoin with Tor is the subject of debate, with recent research suggesting that de-anonymizing Bitcoin users employing Tor is possible given the current manner in which Bitcoin is configured.[30]

The pseudonym process, as mentioned above, does not in itself preserve anonymity; when given access to another user's pseudonym, any individual can see all transactions and balances associated with that pseudonym. Accordingly, Bitcoin recommends changing pseudonyms after every use, though it does not natively enforce this practice.[31] For a VC deployed for everyday use by laypeople, such procedures could be built-in.

In addition, the security-research community has demonstrated the ability to perform privacy-reducing analytics on the overall Bitcoin block chain to attempt to identify individuals solely by the pattern of their transactions.[32] To solve this problem, so-called mixing services exist to obfuscate transactions; these services aggregate transactions so that they cannot be as easily traced to individual actors. Such services

---

[29] See Bitcoin, "Protect Your Privacy," undated (b). For the Tor project, see Tor Project, homepage, undated (c). The security of Tor is outside the scope of this report.

[30] For more details, see Alex Biryukov and Ivan Pustogarov, "Bitcoin over Tor Isn't a Good Idea," paper presented at the 2015 IEEE Symposium on Security and Privacy, San Jose, Calif., May 17–21, 2015a.

[31] See Biryukov and Pustogarov, 2015a.

[32] See, for instance, Sarah Meiklejohn, Marjori Pomarole, Grant Jordan, Kirill Levchenko, Damon McCoy, Geoffrey M. Voelker, and Stefan Savage, "A Fistful of Bitcoins: Characterizing Payments Among Men with No Names," *Proceedings of the 2013 Conference on Internet Measurement IMC '13)*, October 2013, pp. 127–140.

include CoinJoin,[33] Mixcoin,[34] and Dark Wallet,[35] which all seem to provide a sufficient level of user anonymity. Despite this, there is always the threat of future advances in de-anonymization revealing past transactions, even those done with proper anonymity practices.

Having examined anonymity for Bitcoin, we now focus on increased anonymity enabled by some existing and proposed altcoins.

Some New Altcoins Build in Anonymous Transactions

Some altcoins have been built with the primary goal of being more anonymous than Bitcoin. Darkcoin was designed with a mixing service called Darksend built in, which in turn relies on CoinJoin as its underlying technology.[36] All users must participate in the mixing so that de-anonymization is even more difficult—this is an advantage over Bitcoin.[37] Zerocash[38] and its follow-on, Zerocoin,[39] are both built

---

[33] See Bitcoin Forum, "CoinJoin: Bitcoin Privacy for the Real World," discussion thread began August 22, 2013b; and CoinJoin, "Weaknesses in SharedCoin," undated. There are claims, however, that CoinJoin is not as anonymizing as was thought, see for instance http://www.coinjoinsudoku.com. Such research shows the need for more careful analyses of new techniques that are advertised as "privacy enhancing."

[34] See Joseph Bonneau, Arvind Narayanan, Andrew Miller, Jeremy Clark, and Joshua A. Kroll, "Mixcoin: Anonymity for Bitcoin with Accountable Mixes," *Financial Cryptography and Data Security: 18th International Conference*, Berlin: Springer Heidelberg, 2014, pp. 486–504.

[35] See Dark Wallet, undated. There is evidence that terrorists, or their sympathizers, are aware of Dark Wallet; see al-Munthir, 2014.

[36] See Dash, undated (a).

[37] The latest version of Darksend is called Darksend+. For a security evaluation of Darksend+, see Kristov Atlas, "An Analysis of Darkcoin's Blockchain Privacy via Darksend+," self-published article, September 10, 2014; for a response to that work, see Dash Talk, "Reply to Kristov's Paper," self-published article, September 11, 2014.

[38] See Zerocash Project, homepage, undated; Ben-Sasson et al., 2014a; and Ben-Sasson et al., "Zerocash: Decentralized Anonymous Payments from Bitcoin," extended version of the paper presented at the 2014 IEEE Symposium on Security and Privacy, San Jose, Calif., May 18–21, 2014b.

[39] See Zerocoin Project, undated.

using more advanced cryptographic tools.[40] In particular, Zerocash relies on so-called zero-knowledge succinct arguments of knowledge (ZK-SNARKs),[41] which are an advanced cryptographic primitive that obviates the need for some of the distributed-consensus mechanisms of Bitcoin and therefore is able to hide actual transactions to increase anonymity.[42] In other words, the Zerocash approach is to employ advanced cryptographic techniques to anonymize not only *users,* but also *transactions and their patterns.* It is unclear if Zerocash will be adopted; as of the writing of this report, it has not been deployed.

It is difficult to evaluate the ultimate security (and usability) of Zerocash, as it has not been tested in the crucible of real-world use and evaluation, though its theoretical mechanisms have more rigorous security proofs than virtually all existing VCs. By contrast, Dashcoin is currently being used and seems to be reasonably anonymous (certainly more so than using Bitcoin without additional privacy-enhancing technology), though it has only been in existence for about a year as of this writing; its current market capitalization is much less than that of Bitcoin, thus an equal comparison is difficult to make.[43]

So far, we have examined VC anonymity, and in truth, the technological threats we have so far considered have been relatively unsophisticated. In the next section, we will examine cyber threats more

---

[40] There is another altcoin, PinocchioCoin, that is built on top of Zerocoin using different cryptographic techniques; see George Danezis, Cédric Fournet, Markulf Kohlweiss, and Bryan Parno, "Pinocchio Coin: Building Zerocoin from a Succinct Pairing-Based Proof System," *PETShop '13: Proceedings of the First ACM Workshop on Language Support for Privacy-Enhancing Technologies*, New York: Association for Computing Machinery, 2013, pp. 27–30.

[41] See Eli Ben-Sasson, Eli, Alessandro Chiesa, Christina Garman, Matthew Green, Ian Miers, Eran Tromer, and Madars Virza, "SNARKs for C: Verifying Program Executions Succinctly and in Zero Knowledge," in Ram Canetti and Juan A. Garay, eds., *Advances in Cryptology—CRYPTO 2013: 33rd Annual Cryptology Conference*, Santa Barbara, Calif., August 2013, pp. 90–108.

[42] By contrast, Bitcoin relies only on well-understood and accepted (in the computer-security community) hash functions (SHA-256) and digital signature schemes (EC-DSA).

[43] As of February 22, 2015, Bitcoin had a market capitalization of $3,274,674,231, while Dash (at the time, Darkcoin) had a market capitalization of $12,885,950; see CoinMarket-Cap, "Crypto-Currency Market Capitalizations," September 30, 2015a.

broadly, with a particular focus on how the level of cyber sophistication of an adversary might impact the success of a non-state actor VC deployment.

Cyber Threats to Virtual Currencies

One crucial component to examining the possibility of a non-state actor successfully deploying a VC is how much cyber sophistication might be needed to thwart such a deployment. Indeed, if a state actor such as the United States could convince a non-state actor that their VC deployment could be prevented via cyber means, that might shift the decision calculus of the non-state actor away from deployment. There are two related concerns when thinking about how to affect the decision process of non-state actors in their objective to deploy a VC. First, public trust in a currency, as well as the currency's value, could be severely degraded if a VC is compromised via an attack.[44] Second, a VC may be a particularly likely target for the affected nation-state and its allies (including the United States); this is because a VC may be perceived as a national security threat, such as when a VC undermines state-controlled currencies or is used as a means of support for criminals or terrorist groups. Building on this motivation, this section will examine cyber threats to VCs as a function of cyber-threat sophistication.

Ultimately, a non-state actor (indeed, even a state actor) would face significant challenges in protecting a VC from damaging cyber attacks against a determined and sophisticated cyber opponent; the main calculus on the part of the opponent is how much of their capabilities they want to reveal and how much investment in time and personnel they would want in order to ensure a successful attack.

Potential attacks may range from low level (e.g., distributed denial or service [DDoS] attacks) to highly tailored (e.g., attacks against the underlying infrastructure or through the exploitation of zero-day

---

[44] For instance, see Timothy B. Lee, "Major Glitch in Bitcoin Network Sparks Sell-Off; Price Temporarily Falls 23%," *Ars Technica*, March 11, 2013.

vulnerabilities).[45] It should also be noted that attacks may be mounted by opponents other than nation-states; theft of currency, as occurred in the 2014 Bitcoin attack against Mt. Gox,[46] is an obvious motivation.

When discussing cyber threats, it is useful to have a framework for discussing sophistication; we will use the six-tiered system for ranking an actor's sophistication in conducting cyberspace operations as outlined by the Defense Science Board.[47] In particular, Tier I and II opponents are on the level of sophistication of *script kiddies*, that is, individuals or groups who use commonly available exploits. Tier III and IV opponents are more sophisticated, developing their own custom malicious code (e.g., based on zero-day vulnerabilities, which may have been discovered by the opponents themselves, or orchestration of multiple-attack vectors and vulnerability exploitation).[48] Tier V and VI opponents, while capable of sophisticated Internet-based attacks, will work to create vulnerabilities and opportunities for attack. Tier V and VI opponents will use not only sophisticated cyber techniques but also sophisticated human intelligence (HUMINT) capabilities as well; in this sense they are truly full-scope actors. For more detail about the tiers, including further examples, see Chapter Two of the Defense Science Board's *Task Force Report: Resilient Military Systems and the Advanced Cyber Threat*; see the appendix of this report for a table describing the tiers.

---

[45] A *zero-day exploit* is an attack that takes advantage of a software vulnerability that the developer is unaware of and for which no patch exists. While this section will discuss some attacks, see also Bonneau et al., 2015; Atlas, 2014; and Bitcoin Wiki, "Weaknesses," July 8, 2015d, for other detailed discussions of particular attacks as well as potential countermeasures.

[46] Details concerning the theft of nearly $400 million from Mt. Gox are still emerging, and there is some dispute about the causes of the loss. Although the company was apparently mismanaged, it seems clear that a combination of insider support of hacker attacks resulted in the losses. See Robert McMillan, "The Inside Story of Mt. Gox, Bitcoin's $460 Million Disaster," *Wired* online, March 3, 2014.

[47] Defense Science Board, Department of Defense, *Task Force Report: Resilient Military Systems and the Advanced Cyber Threat*, January 2013.

[48] For instance, the malware Flame was the work of Tier IV opponents, according to the Defense Science Board, 2013.

Importantly, high-tier opponents could rely on lower-tier techniques, and often do so to obfuscate their identities and their capabilities. In particular, an attacking opponent may be unwilling to use high-tier techniques not because of their sophistication, but rather due to an unwillingness to have the attack be attributed to them.[49] A related issue is cost-benefit analysis related to whether the investment in time and money standing up a new cyber capability is worth successfully degrading/destroying a VC using cyber means.[50] In this section, we will call adversaries of the non-state actor deploying a VC *opponents*; as mentioned above, these opponents may include both the nation-state(s) where a VC is deployed as well as allies of that "victim" nation-state, who may have far more advanced cyber capabilities.

It should be noted at the outset that the easiest, most effective attacks that the nation-state where a VC is deployed can undertake is either shutting off or strongly filtering the Internet of a country where VC transactions are originating from; such attacks would be particularly effective for denying access to digital-wallet services and mining services. Shutting off the Internet entirely, however comes with serious costs, while effectively filtering the Internet for a country that is not already doing so could require significant additional resources. In addition, any firewall could be defeated by sufficiently effective IP-masking techniques such as Tor, though in practice, such techniques would have to be built in to a VC software.[51] Finally, such filtering would not

---

[49] The decision of a state actor to use a particular cyber capability, or reveal a particularly sophisticated technological approach to cyberspace operations, is beyond the scope of this report.

[50] Another consideration of a nation-state may be political backlash arising from denying a population access to their currency, thereby achieving the technical goal of destroying/degrading a VC at the possible cost of losing heart and minds. This is an interesting avenue for future study.

[51] It is unclear if Tor can work in the presence of a well-implemented firewall. For instance, Tor has encountered challenges against China's Great Firewall. See, for instance, Roya Ensafi, Philipp Winter, Abdullah Mueen, and Jedidiah R. Crandall, "Large-Scale Spatio-temporal Characterization of Inconsistencies in the World's Largest Firewall," self-published paper, October 3, 2014, and Philipp Winter and Stefan Lindskog, "How the Great Firewall of China is Blocking Tor," paper presented at the Second USENIX Workshop on Free and Open Communication on the Internet (FOCI), Bellevue, Wash., August 2012.

entirely prevent the employment of VCs locally, provided local Internet infrastructure and computing power supports the sustainment of the currency.

Finally, we distinguish *attack path* from the vulnerability exploitation and attack itself. Attack paths can be thought of as the way in, whether through spear phishing, backdoors, deliberate implant by a human (e.g., human agent delivery of malware through a USB flash drive), or through the propagation of viruses by way of mobile devices. Vulnerability exploitation is the use of a computer or network's features for additional purposes, whether through, for example, DDoS attacks (exploiting the properties of digital networks), buffer overflows (overwriting memory to implant code), or tampering.[52]

Attacks Used by Tier I and Tier II Opponents

Tier I and Tier II opponents have a variety of potential attacks they could perform. The most straightforward is an attack that relies on straight computational power or bandwidth.[53] Indeed, perhaps one of the most powerful attacks they could perform against decentralized VCs such as Bitcoin is to exceed the computational strength that would otherwise secure the system (typically 51 percent of the total computational power, also often called *mining power*).[54] The most powerful

---

[52] A recent example of tampering occurred with the sale of Lenovo personal computers, on which the company installed adware that embeds itself in the computer registry and supplies fraudulent security certificates to websites. Since VCs rely on certification and verification of ownership, this vulnerability could be used to compromise a system and its stored cryptocurrency keys.

[53] It is worth nothing that, while attacks on computational power are straightforward against Bitcoin, alternative models such as Proof-of-Stake and its derivatives are not susceptible in the same way, as they do not rely on computing power for generation of the block chain. In addition, the monetary expense of these attacks may be large when conducted by a single actor.

[54] The 51 percent number is much talked about in practice, though there are theoretical results that show the limit may be closer to 25 percent; see Ittay Eyal and Emin Gun Sirer, "Majority Is Not Enough: Bitcoin Mining Is Vulnerable," in Nicolas Christin and Reihaneh Safavi-Naini, eds., *Financial Cryptography and Data Security: 18th International Conference, FC 2014*, March 2014, pp. 436–454. In addition, Juan Garay, Aggelos Kiayias, and Nikos Leonardos, "The Bitcoin Backbone Protocol: Analysis and Applications," in Elisabeth Oswald and Marc Fischlin, eds., *Advances in Cryptology—EUROCRYPT 2015: 34th Annual*

of such attack to destroy a currency is a so-called Goldfinger attack.[55] Indeed, as Bonneau and coauthors point out:

> If a majority miner's goal is explicitly to destroy Bitcoin's stability and hence its utility as a currency, they can certainly do so by intentionally introducing deep forks or rejecting other miner's blocks . . . *A state wishing to damage Bitcoin to avoid competition with its own currency,* [emphasis added] or an individual heavily invested in a competing currency, may be motivated to attempt such an attack.[56]

In other words, a Goldfinger attack comprises a cartel formation, in which the cartel, through its dominant computational power, can change the market rules (to undermine faith in the currency), disallow certain users of the currency (to drive out a subset of users from the currency market), or strangle new currency supplies (to drive up prices). Unless the currency was new or the amount of computational power associated with decentralization was otherwise low, it is unclear how an opponent would be able to gain persistent access to 51 percent of the total computational power.[57]

Particularly for Bitcoin, there is another avenue to perform a Goldfinger attack, namely through corrupting the mining pools. Mining is typically performed by computational pools that work by aggregating the mining effort of individual miners. Some of these mining pools

International Conference on the Theory and Applications of Cryptographic Techniques, April 2015, pp. 281–310, demonstrate that the 51 percent number also degrades in the presence of increased network latency.

[55] See Joshua A. Kroll, Ian C. Davey, and Edward W. Felten, "The Economics of Bitcoin Mining or, Bitcoin in the Presence of Adversaries," paper presented at the 12th Workshop on the Economics of Information Security (WEIS 2013), Washington, D.C., June 11–12, 2013.

[56] Bonneau et al., 2015, continue: "Arguably, these attacks have already been observed through altcoin infanticide, in which deep-forking attacks against new competing currencies with low mining capacity have been successfully mounted by Bitcoin miners."

[57] One way to capture 51 percent of the computational power could be either through cryptographic breakthroughs (e.g., algorithmic methods to break SHA-2 faster in the case of Bitcoin) or through novel computing hardware, from special-purpose circuits to possibly quantum computers. The possibility of these occurrences is outside the scope of this report.

can approach the 51 percent threshold, including the notable case of GHash.io, which briefly exceeded the threshold and then promised not to do it again.[58] The issue here is not that a mining pool might decide to crash Bitcoin; rather the issue is that an attacker could attempt to hack several mining pools that would then correspond to greater than 51 percent computational power. In such a manner, an attacker with relatively little initial resources could mount a 51 percent attack on Bitcoin. In practice, such an attack may require a high-tiered opponent.

It should be noted that such an attack requires a capital investment outside of the Bitcoin market (for computers and electricity, for instance), so it can be difficult to calculate the return on investment for such an attack. Nevertheless, if the cost is within the bounds of cost that the opponent might otherwise spend on weaponry for a direct (kinetic) attack against the currency sponsors, a Goldfinger-type attack should be thought of as realistic.[59]

In the case of centralized or semi-centralized VCs, DDoS attacks and spear phishing to attack vulnerabilities in the networking and computational infrastructure may be effective in degrading a VC system, particularly at more centralized services such as online wallets or mining services. A set of related DDoS attacks exist that include transaction spamming and script attacks to waste computing power by creating transactions that need significant computation to verify.[60]

These are other ways an attacker can impose costs on the network, even if there is no central authority. Attacking exchanges or other more centralized cyber services may prove effective, even if a VC is decentralized. Rather low-tech methods may even be used to attack Bitcoin users using Tor.[61] DDoS attacks can be used to degrade general net-

---

[58] See Bonneau et al., 2015, and GHash.io, "Bitcoin Mining Pool GHash.IO Is Preventing Accumulation of 51 Percent of All Hashing Power," undated.

[59] Another consideration may be the cost of legal action such as arrests to thwart the currency, which has been successful against VCs such as Liberty Reserve. See *United States v. Liberty Reserve*, 2013, and Department of Justice, 2013.

[60] See, for instance, Bitcoin Forum, "New Bitcoin Vulnerability: A Transaction That Takes at Least 3 Minutes to Be Verified by a Peer," discussion thread began January 30, 2013a.

[61] See Biryukov and Pustogarov, 2015a.

work connectivity for local, everyday economic transactions so that VC transactions are too slow to be practical or convenient. Any attack that compromises systems that have access to the keys for user accounts, or that compromises the users systems,[62] can be used to steal currency.

It should be noted that the vast majority of existing literature dedicated to VC security seems to be relative to Tier I and II threats, which is understandable since such low-level threats are already rather effective. We will now consider more advanced threats.

Attacks Used by Tier III and Tier IV Opponents

Tier III and Tier IV opponents would employ more sophisticated attacks including discovery and exploitation of zero-day vulnerabilities or manipulating the underlying VC infrastructure. For instance, in Bitcoin, "How participants in the Bitcoin ecosystem achieve consensus about the default rules for Bitcoin transactions is under-analyzed."[63] Since the Bitcoin system requires user consensus on rules for currency generation and transaction state and its validation, it is susceptible to manipulation of those rules, or exploitation of gaps or flaws in the rule implementation. Indeed, high-tier opponents may look to attack the underlying rules of decentralized VCs to change them.

Tier III and IV opponents also have the capability to discover and exploit zero-day attacks and may use them to great effect. In particular, they may use them to attack the mining pools, as discussed in the previous section above, in order to gain control of 51 percent of total computational power. In the centralized and semi-centralized authority cases, such actors may successfully exploit authority servers and essentially crash the currency, perhaps through an orchestrated attack that involves insiders. Even in the decentralized case, advanced opponents can successfully exploit specific targets with high probability and can

---

[62] The wide variety of extant exploits that are exploitable by hackers with minimal skills makes securing each user computer or smartphone critical. Viruses that steal bitcoin already exist and have been publicly observed. See Adrian Covert, "There's a Virus That Will Steal All Your Bitcoins," *Gizmodo.com*, June 17, 2011. This again points to the importance of tradecraft for users of crypto-currencies.

[63] Bonneau et al., 2015.

publically target high-net-worth individuals to reduce confidence in the currency (or can randomly target average citizens to sow distrust).

Tier IV opponents would likely have the capability to construct and use zero-day exploits against critical VC services such as exchanges and wallets as well as cell-phone applications used to conduct everyday transactions. Indeed, they may look to use fake permissions and certificates to install applications that subvert (or spy on) user VC applications. They would then either disrupt those applications or publicize vulnerabilities to degrade confidence in a VC. Tier IV opponents might also attempt to degrade the ability of a VC system to construct reliable cryptographic protocols (such as key generation and storage as used by wallet applications) by subtly changing the software implementations of key cryptographic functionalities. They may attempt to change the actual code used by VC servers or users in order to degrade functionality or allow for an easier attack path to later simultaneously deny service to broad classes of servers and/or users.

Attacks Used by Tier V and Tier VI Opponents
Tier V and Tier VI actors could employ particularly damaging attacks through supply-chain attacks against the underlying infrastructure or through subverting the implementation of the software used by VC participants. These actors may infect broad classes of software and hardware. They might target cell phones or other hardware, including computers used as servers for critical VC services or special-purpose hardware used for mining, and corrupt them before delivery. They could leverage this access to enable them to conduct the operations listed in the above section on Tier III and IV actors with a higher probability of success. By infecting hardware and in particular the special-purpose hardware that performs cryptographic tasks, Tier V and VI actors may also be able to break cryptographic standards that underlie the security assumptions of a VC, which could in turn completely break the security of a VC. If publicly revealed (or revealing the consequences of such a break without revealing the break itself), this strategy could result in a severe degradation of confidence in a VC.

Tier V and VI actors could also employ HUMINT methods, namely by employing agents to assume the roles of key VC personnel, such as software developers, or to bribe or otherwise co-opt such personnel, either within a VC's organization or at other organizations that provide critical services to a VC.

The Possibility of Successful Defense

In light of the previous discussion on attacking a VC, it is worth briefly examining whether it is possible *at all* to deploy a VC that could withstand a cyber attack. Otherwise, discussing non-state actor deployment of a VC would be moot.

Ultimately, it seems clear that a non-state actor (indeed, even a state actor) would face significant challenges against a determined high-tiered opponent given the underlying assumptions and implementation of VCs. As a general matter, a high-tiered opponent would be able to successfully attack any target of interest in cyberspace if enough resources were invested. In the case of a VC, which would require trust, anonymity, and availability of widely deployed cyber services (such as wallet and mining applications), it seems infeasible that a consistently successful cyber defense can be mounted. The only hope might be if the non-state actor were supported by a sophisticated nation-state opponent who was capable of defending against such threats. Even in this scenario, it is unclear whether such coordination would work, particularly in the case of a Tier V and VI opponent.

Against a Tier I and II opponent, a sophisticated non-state actor may be able to mount a defense, though protecting against DDoS or mining attacks may be tied to how centralized various services are and how much mining power is currently supporting decentralized operations.

It is worth mentioning again that high-tiered opponents may not wish to conduct attacks that reveal their sophistication. Indeed, political calculations may prevent a high-tiered opponent from mounting their most sophisticated attack, particularly in circumstances in which a non-state actor has shown the capability to retaliate or when a non-state actor is backed by a nation-state that might view such a sophisticated attack as enough of a threat to retaliate. In sum, the great unknown is not necessarily whether or not a sophisticated opponent would be capable of bringing down a VC, but whether they would be willing to politically or have the capacity to dedicate the resources to a VC as a prioritized target.

It might be possible, however, to devise a stable and relatively resilient VC by revisiting some assumptions and the underlying architectures, particularly when those assumptions and architectures are articulated by a sophisticated cyber power (such as a nation-state's cyber subject-matter experts). In particular, it is crucial that proper measures are taken to protect the design of the underlying VC software as well as external services, such as secure wallet services (e.g., smartphone applications) and mining services, together with the associated protection of the servers that would run these services. A sophisticated nation-state is the most capable actor to ensure this security, which is another reason why a VC has the greatest chance for cyber survival when a non-state actor is supported by a nation-state that possesses cyber sophistication. At the very least, the level of sophistication and investment to successfully attack a VC would be raised, making any opponent's decision calculus to attack a VC more complicated.

CHAPTER FIVE

Implications Beyond Currency

Previous chapters in this report examined the potential challenges associated with the deployment of VCs by non-state actors as currency. We now examine the broader technological implications of VC development, particularly in the context of national security. While there has been much public discussion about applications of the Bitcoin block-chain technology for future finance-related technology advances,[1] our interests in this subject are broader and include: (1) direct implications of Bitcoin-style block-chain technology, (2) how VC development might result in greater cryptographic sophistication of previously unsophisticated actors, and (3) how VCs may spell the beginning of an era in which low-sophistication cyber actors have easy access to resilient cyber services.

As we discuss in this chapter, the ultimate national-security policy concern is the availability of increasingly sophisticated cyber capabilities to low-sophistication cyber actors. While such availability could benefit the United States by providing U.S.-supported non-state actors access to greater cyber services in potentially denied or degraded cyber environments, it could also harm U.S. interests by allowing terrorist groups access to cyber services that would prove increasingly difficult for the country to thwart.

---

[1] See, for instance Marc Andreessen, "Why Bitcoin Matters," *New York Times* online, January 21, 2014.

Block-Chain Technology and Distributed Consensus

Ultimately, decentralized VCs such as Bitcoin have provided a resilient means to store and update data in a highly distributed fashion that is very hard to corrupt. In the case of Bitcoin, their data is a public-transaction record, but in principle, other data could be stored in an analogous fashion. The time required to distribute and agree upon the data is also a limiting factor. Even for VCs that are more efficient than Bitcoin, the consensus time is on the order of minutes. This lag time seems inherent in decentralized systems, where many nodes must agree on a common operational state. Another major issue is the communication bandwidth required; as data requirements of the cyber service increase, the amount of communication throughout the decentralized network also increases. A question for future research would thus be how to perform such services in a communication-efficient fashion.

In general, possible use of such technology arises from situations where the broad dissemination and maintenance of data would otherwise be challenged or suppressed, but access to the data is not on the order of milliseconds. Examples include tactics, techniques, and procedures (TTPs) designed to enable political dissidents and the dissemination of terrorist publications, such as al-Qaeda in the Arabian Peninsula's (AQAP) *Inspire* or ISIL's *Dabiq*.

One of the central challenges in adapting block-chain technology to other nonfinancial applications will be how to incentivize the securing of a decentralized system. One of Bitcoin's main innovations is how well it intertwines economic incentives with a decentralized security process. For example, there has been work on increasing the security of the decentralized Tor service, but by using financial incentives.[2] It is unclear how Bitcoin-style block-chain technology will progress in settings without such incentives.[3]

---

[2]   See Alex Biryukov and Ivan Pustogarov, "Proof-of-Work as Anonymous Micropayment: Rewarding a Tor Relay," paper presented at the 19th International Conference on Financial Cryptography and Data Security 2015, San Jose, Puerto Rico, January 26–30, 2015b.

[3]   For one possible solution, see Maidsafe, undated (a).

As the efficiency of secure, resilient distributed consensus increases (and communication requirements are kept manageable), more time-sensitive tasks can occur. These include resilient online forums that might be difficult to disrupt, real-time resilient chat services, and anonymous messaging services routed throughout large-scale decentralized networks, creating true anonymity.[4] At their most efficient, distributed consensus protocols could be used for resilient, anonymous voice-over-Internet-protocol (VOIP) conversations—a truly resilient and anonymous version of Skype.

While such applications could be useful for national-security users, they would be particularly beneficial for adversaries with low technological sophistication, since they would be able to access far more resilient services than they would have otherwise considering their limited skill set.

Virtual Currencies Increase Cryptographic Sophistication

Increased awareness of block-chain technologies has, as a result, increased awareness of sophisticated cryptographic techniques for distributed consensus and computation. Venture capitalists now talk about computer-science concepts such as Byzantine Generals Problem,[5] and general cybersecurity experts now talk about deep results in theoretical cryptography such as ZK-SNARKs.[6] Ordinarily, these topics would never have been the subject of discussion beyond rarefied academic circles.

One possible outcome is a greater focus on the applications of advanced cryptographic techniques such as secure multiparty computation (MPC): the field of cryptography seeking to perform distributed computation in a manner that preserves the confidentiality and

---

[4] Jonathan Warren, "Bitmessage: A Peer-to-Peer Message Authentication and Delivery System," self-published paper, November 27, 2012.

[5] See Andreessen, 2014. Unfortunately, the claim about Byzantine Agreement in the article is incorrect (see Garay, Kiayias, and Leonardos, 2015).

[6] See Ben-Sasson et al., 2013.

integrity of computation inputs and outputs, even in the presence of malicious activity within the distributed system. The distributed consensus Bitcoin protocol is, in a sense, a special set of functionalities that MPC attempts to compute.[7] Increased focus on MPC could yield evermore efficient and secure distributed protocols for ever-increasing classes of functionality; the Defense Advanced Research Projects Agency (DARPA) recently demonstrated a secure VOIP application over untrusted infrastructure using MPC.[8]

Another outcome is the increasing availability of well-designed cryptographic software—or, more generally, code—that, originally designed to support VCs, which can now be used by less sophisticated software developers to enable greater security. In practice, this might allow cyber criminals and terrorists with a lower level of technological sophistication to afford more secure communications and other cyber services, making it increasingly difficult for the U.S. government to track and defeat them.

Finally, increased mining-based VC use might have implications for the availability of special-purpose hardware to break cryptographic security. For example, the process of mining Bitcoin is the same process employed to crack the SHA-2 cryptographic hash function. Currently, hardware miners are capable of performing over 5 trillion hashes per second; to put this in perspective, only 1,000 of these miners would have accounted for the *total* mining power of Bitcoin in December 2013, at the height the VC's market capitalization.[9] The economic

---

[7]   Bitcoin-style consensus is not exactly a special case of MPC. Bitcoin-style decentralization is particularly interesting because it *incentivizes* distributed computation, whereas MPC typically assumes no such incentivization. In a sign of how much Bitcoin is on the minds of those interested in MPC (and vice versa), the best paper prize at the 2014 IEEE Symposium on Security and Privacy was titled "Secure Multiparty Computations on Bitcoin" (see Marcin Andrychowicz, Stefan Dziembowski, Daniel Malinowski, and Łukasz Mazurek, "Secure Multiparty Computations on Bitcoin," paper presented at the IEEE Symposium on Security and Privacy, San Jose, Calif., May 18–21, 2014).

[8]   See Defense Advanced Research Projects Agency, "DARPA I2O Demo Day Featured Programs," May 21, 2014.

[9]   See Bitcoin Wiki, "Mining Hardware Comparison," September 16, 2015c, and Blockchain, "Market Capitalization," undated (b).

incentivization toward evermore powerful hardware that could break cryptographic security may rival nation-state investments in similar hardware, which could have broad implications for the security of cryptographic tools.

Virtual Currencies and the Trend Toward Resilient, Decentralized Cyber Services

Bitcoin and current innovations in VCs can be seen as merely the latest step toward giving low-sophistication cyber actors access to decentralized, resilient cyber services. Understanding this historical trend can help define where the trend may be headed and what implications there may be for the Department of Defense (DoD).

The first step toward the development of Bitcoin were peer-to-peer technologies such as Napster and Gnutella (and later, BitTorent). These technologies allowed users to access information by connecting with strangers on the Internet, thus providing a forum to exchange data.[10] These services revolutionized the availability of data and had a large impact on entities such as the music industry. The overall security of these services was minimal. Since transactions were bilateral, those publicly offering particular cyber services (in this case, data) could be disrupted and monitored with relative ease. This phase of technology could be seen as the "cyber availability without decentralization" phase.

The first move toward decentralization came with the Tor project.[11] Tor allows users to keep their Internet identities private by providing a pool of available nodes around the world that a user can then access sequentially. From a monitor's perspective, the user appears to

---

[10] See Johan Pouwelse, Paweł Garbacki, Dick Epema, and Henk Sips, "The Bittorrent P2P File-Sharing System: Measurements and Analysis," in Miguel Castro, ed., *IPTPS 2005 Proceedings of the Fourth International Conference on Peer-to-Peer Systems*, February 2005, pp. 205–216, and Stefan Saroiu, P. Krishna Gummadi, and Steven D. Gribble, "A Measurement Study of Peer-to-Peer File Sharing Systems," Martin G. Kienzle and Prashant J. Shenoy, eds., *Proceedings of SPIE: Multimedia Computing and Networking (MMCN) 2002*, Vol. 4673, 2002, pp. 156–170.

[11] For more information about Tor, see Tor Project, "Overview," undated (d).

have the identity of the last node used in the Tor service. Tor is made possible because volunteers around the world host the nodes that other users can use. In addition, entire websites can be created, which are only accessible via Tor. This system, in aggregate, is often called the Dark Web.[12] Tor is much harder to attack since users hop from Tor node to Tor node, but the number of hops is small (typically three), and Tor is thought to be vulnerable to sophisticated adversaries.[13] From a centrality perspective, Tor can be called *loosely decentralized*. From a wide group of available Tor nodes, a user ends up relying on very few of them, but has a choice of which nodes to use.

Bitcoin and other decentralized VCs move toward *full* decentralization, where the cyber service—distributed consensus—actually relies on the *majority* of the decentralized network rather than a small number of nodes (in the case of Tor). As addressed in Chapter Four, it is possible to successfully attack VCs. By moving toward ever-more decentralized cyber operations, however, relatively unsophisticated cyber actors are better able to have easy access to increasingly sophisticated cyber services. The national-security community will have to contend with the challenge of thwarting these cyber actors over the coming years.

Toward Resilient, Public Cyber Key Terrain

About ten years passed between the creation of Napster and the creation of Bitcoin. This rapid pace of development begs the question: "What technologies do we anticipate in the next ten to 20 years?" As a thought experiment, the historical trend poses the idea of *resilient, public cyber key terrain*: the ability for unsophisticated cyber actors to have persistent, assured access to cyber services *regardless* of whether a

---

[12] See Tor Project, "Hidden Service Protocol," undated (b), and "Anonymity Online," undated (a).

[13] For an example of one such attack, see Tor Project, "Security Advisory Relay Early Traffic," July 30, 2014, and "Category, Tags, Attacks," December 19, 2014.

highly sophisticated state actor opposes their use. What are the technical implications of such availability?

Overall, resilient public cyber key terrain could prove a double-edged sword: enabling DoD to project power, both in terms of information as well as cyberspace operations, but also enabling enemies of the United States to do the same, and with a lower barrier of entry than before.

Resilient public cyber key terrain may enable the truly free flow of information in the form of public communications, such as uninterruptible news sites and web forums, breaking down national firewalls, such as China's Great Firewall, but also enable even greater access to extremist rhetoric and tactics. Such a capability would defeat Internet censorship and enable the projection of the American view of the world into countries that previously denied such information. At the same time, crime- and terrorism-enabling websites would be a permanent fixture (indeed, in such an environment, the website for Liberty Reserve, a precursor VC that was shuttered by the U.S. government, might still be operational). One possible response to such advances may be for some countries to fundamentally separate themselves from the global Internet. DoD would both be able to conduct information operations with greater freedom but would also be more susceptible to terrorist information operations.

Direct access to resilient cyber services may enable a global, resilient communication infrastructure that would enable private communication: uninterruptible, anonymous, and encrypted communication.[14] Such links could serve the communication needs of political dissidents to communicate without interference from their government. It could, however, also enable criminals, terrorists, or even nation-states looking to set up unattributable cyber infrastructure with which to plan cyber attacks or conduct criminal activity. The implications for DoD and the

---

[14] More technically, it might be used for a global public key infrastructure (PKI). A PKI is a means to enable a root of trust for secure communications and authentication for users and devices. For more about PKI, see, for instance, Richard D. Kuhn, Vincent C. Hu, W. Timothy Polk, and Shu-Jen Chang, *Introduction to Public Key Technology and the Federal PKI Infrastructure*, Gaithersburg, Md.: National Institute of Standards and Technology, U.S. Department of Commerce, February 26, 2001.

intelligence community may be that entirely new signals intelligence TTPs would have to be developed to defeat this threat.

Beyond using resilient cyber key terrain for communication, adversaries of the United States may use such technologies to enable unhackable and unattributable hop points for nation-state cyber attacks. Actors, both state and non-state, may be able to project cyber power in ways that were never before possible and at far lower levels of cyber sophistication than were possible before. On the other hand, DoD could use the same resilient infrastructure to gain access and conduct cyber attacks as well. One possible implication may be an entirely new push for more effective cyber defenses than what currently exists.

CHAPTER SIX

Conclusions and Future Research

This report examined the potential for terrorist, insurgent, or criminal groups to increase their political and/or economic power by *deploying* a VC to use as a currency for regular economic transactions. To thwart the threat of non-state VC use, the national-security community of the United States should understand how these non-state actors might exploit VCs. This examination is a small part of a larger conversation on the feasibility of VCs, both from a social-science perspective (i.e., VC as currency) as well as a technological perspective (i.e., VC as secure, anonymous, and resilient cyber service).

From an economic perspective, promoting adoption of VCs (versus adopting established currencies) may face significant challenges of acceptance by the population, both as a new currency with no previous history and thus potentially lacking in legitimacy and as a currency that has no tangible representation in the form of paper and coins in societies accustomed to conceiving of money in terms of its physical manifestations. We expect populations' suspicions of VCs will erode as they become more familiar with them. Changes in attitude could alter this as the technologies that underlie VCs become more prevalent and trusted. Moreover, in a territory in which a VC is the only medium of exchange, economic necessity may force people to accept VCs where they would have otherwise rejected them.

Consequently, perhaps the best strategy for the United States and its allies to thwart a VC deployment would be to target those properties of a VC that would most increase its acceptance, most notably transaction anonymity, security, and availability.

From a technological perspective, deploying a VC as a replacement currency for everyday economic transactions is very challenging today. Challenges include having access to the technological sophistication necessary to develop, deploy, and maintain a VC as a cyber service; ensuring levels of transaction anonymity demanded by users while ensuring transaction integrity so that buyers and sellers are assured of proper exchange, all without the need for overly advanced technological expertise; and finally, protecting the overall integrity (and availability) of a VC against advanced cyber threats, particularly those nation-states that would oppose the non-state actor's VC deployment. The availability of technologies to mitigate these issues in the future could make it far more feasible for a non-state actor to deploy a VC.

At the same time, an important decision calculus for the United States or others looking to thwart a VC is what level of sophistication (or capabilities) are worth demonstrating, and what kinds of investments in people, time, and research are worthwhile in order to thwart a VC.

The deployment of a VC by non-state actor is most feasible when supported by a nation-state with advanced cyber expertise. This nation-state could enable the non-state actors to overcome the considerable technical hurdles associated with deploying a VC. Included here is the ability of a nation-state to defend a non-state actor by a sophisticated cyber attack from another nation-state opponent, such as the United States.

Finally, the development of VCs have the potential to advance technologies of relevance to the national-security community, including increasingly resilient cyber services to low-sophistication actors, such as information dissemination and storage. Increased focus on VCs has also increased cryptographic sophistication, which may lead to increasingly secure software on the one hand and increasingly efficient hardware for breaking cryptographic security on the other. Finally, the historical trend suggests the development of a resilient public cyber key terrain, which this report defined as the ability of unsophisticated cyber actors to have persistent, assured access to cyber services regardless of whether a highly sophisticated state actor opposes their use. This has implications for government-imposed firewalls to censor Internet content; access to extremist rhetoric; the feasibility of nation-state cyber attacks; and the ability to maintain secure, uninterruptible, and anonymous communication.

For Future Research

There are many challenges arising from this report that deserve future study. One is gaining a clearer understanding of the tradeoffs between technological barriers to entry for VCs and the willingness of a population that uses them for everyday transactions. A usability study would be particularly beneficial. A related challenge is understanding the degree to which the average citizen is willing to trust VCs based on advanced cryptographic principles that they do not comprehend. This would have implications for new currencies, such as Zerocash, that have been touted as a significant step forward for anonymity.

Significantly more work is needed to examine the potential for non-state actor exploitation, rather than deployment, of a VC. When might a non-state actor choose to use a VC rather than cash for illicit transfer, fundraising, or money laundering? What would the key decision points be? Which VCs are most feasible for use in this way? What total volume of currency is feasible for regular, or semiregular, transfer while still maintaining anonymity? In particular, when would a VC be more useful for illicit transfer than physical U.S. dollars? This report may be used to inform this examination, however, much more work needs to be done.

Closer examination is needed of regions around the world and non-state actors that would most benefit from deploying a VC as a political tool. Additional analysis is needed for what indicators and warnings would be most useful for demonstrating that non-state actors are increasingly relying on VCs.

Further study is needed to understand what resilient technologies VCs do and do not enable. It is unclear whether the economic incentives that enable Bitcoin would work for more general cyber services. Finally, further study is needed to explore the ideas behind resilient, public cyber key terrain, both for long- and near-term policy implications, as increasingly resilient cyber services become widespread and available for even the least sophisticated cyber user.

Rating Cyber Threat Sophistication by Tiers

Below is a summary of the tiered framework as outlined by the Defense Science Board in table contained in their document from 2013. See Chapter Two for more details.

Table A.1
Cyber Threat Tiers

Tier	Description
I	Practitioners who rely on others to develop the malicious code, delivery mechanisms, and execution strategy (use known exploits)
II	Practitioners with a greater depth of experience and an ability to develop their own tools (from publicly known vulnerabilities)
III	Practitioners who focus on the discovery and use of unknown malicious code, are adept at installing user and kernel mode-root kits, frequently use data-mining tools, target corporate executives and key users (government and industry) for the purpose of stealing personal and corporate data with the expressed purpose of selling the information to other criminal elements
IV	Criminal or state actors who are organized, highly technical, proficient, well-funded professionals working in teams to discover new vulnerabilities and develop exploits
V	State actors who create vulnerabilities through an active program to "influence" commercial products and services during design, development, or manufacturing, or with the ability to impact products while in the supply chain to enable exploitation of networks and systems of interest
VI	States with the ability to successfully execute full-spectrum (cyber capabilities in combination with all of their military and intelligence capabilities) operations to achieve a specific outcome in political, military, and economic domains and apply at scale

SOURCE: Defense Science Board, 2013.

References

al-Munthir, Taqi'ul-Deen, "Bitcoin wa Sadaqat al-Jihad: Bitcoin and the Charity of Violent Physical Struggle," self-published article, August 2014. As of February 26, 2015:
https://alkhilafaharidat.files.wordpress.com/2014/07/btcedit-21.pdf

Altcoins, homepage, undated. As of February 24, 2015:
http://altcoins.com

Andreessen, Marc, "Why Bitcoin Matters," *New York Times* online, January 21, 2014. As of February 23, 2015:
http://dealbook.nytimes.com/2014/01/21/why-bitcoin-matters

Andrychowicz, Marcin, Stefan Dziembowski, Daniel Malinowski, and Łukasz Mazurek, "Secure Multiparty Computations on Bitcoin," paper presented at the IEEE Symposium on Security and Privacy, San Jose, Calif., May 18–21, 2014.

Apple, "iOS Security, iOS 9.0 and Later," September 2015. As of October 7, 2015:
https://www.apple.com/business/docs/iOS_Security_Guide.pdf

Atlas, Kristov, "An Analysis of Darkcoin's Blockchain Privacy via Darksend+," self-published article, September 10, 2014. As of February 22, 2014:
http://cdn.anonymousbitcoinbook.com/darkcoin/darksend-paper/Atlas_Darksend-Analysis-v001.pdf

Auroracoin, "AuroraSpjall" undated. As of November 6, 2015:
http://auroraspjall.is

Barotseland Free State, *Barotseland Mupu Currency Act of 2012*, February 28, 2012. As of April 16, 2015:
http://www.barotseland.info/Currency_Act.htm

Ben-Sasson, Eli, Alessandro Chiesa, Christina Garman, Matthew Green, Ian Miers, Eran Tromer, and Madars Virza, "SNARKs for C: Verifying Program Executions Succinctly and in Zero Knowledge," in Ram Canetti and Juan A. Garay, eds., *Advances in Cryptology—CRYPTO 2013: 33rd Annual Cryptology Conference*, Santa Barbara, Calif., August 2013, pp. 90–108.

————, "Zerocash: Decentralized Anonymous Payments from Bitcoin," paper presented at the 2014 IEEE Symposium on Security and Privacy, San Jose, Calif., May 18–21, 2014a.

————, "Zerocash: Decentralized Anonymous Payments from Bitcoin," extended version of the paper presented at the 2014 IEEE Symposium on Security and Privacy, San Jose, Calif., May 18–21, 2014b. As of February 20, 2015: http://zerocash-project.org/media/pdf/zerocash-extended-20140518.pdf

Bernstein, Peter L., *The Power of Gold: The History of an Obsession*, Hoboken, N.J.: Wiley and Sons, Inc., 2004.

Biryukov, Alex, and Ivan Pustogarov, "Bitcoin over Tor Isn't a Good Idea," paper presented at the 2015 IEEE Symposium on Security and Privacy, San Jose, Calif., May 17–21, 2015a.

————, "Proof-of-Work as Anonymous Micropayment: Rewarding a Tor Relay," paper presented at the 19th International Conference on Financial Cryptography and Data Security 2015, San Jose, Puerto Rico, January 26–30, 2015b.

Bitcoin, "Choose Your Bitcoin Wallet," undated (a). As of February 19, 2015: https://bitcoin.org/en/choose-your-wallet

————, "Protect Your Privacy," undated (b). As of February 22, 2015: https://bitcoin.org/en/protect-your-privacy

————, "Some Things You Need to Know," undated (c). As of February 20, 2015: https://bitcoin.org/en/you-need-to-know

Bitcoin Forum, "[RELEASE] Liquidcoin (Speculation Based)," discussion thread began January 18, 2012. As of February 26, 2015: https://bitcointalk.org/index.php?topic=60026.0

————, "New Bitcoin Vulnerability: A Transaction That Takes at Least 3 Minutes to Be Verified by a Peer," discussion thread began January 30, 2013a. As of October 13, 2015: https://bitcointalk.org/index.php?topic=140078.msg1491085#msg1491085

————, "CoinJoin: Bitcoin Privacy for the Real World," discussion thread began August 22, 2013b. As of February 22, 2015: https://bitcointalk.org/index.php?topic=279249.0

Bitcoin Help, homepage, undated. As of February 25, 2015: https://bitcoinhelp.net

Bitcoin Wiki, "Comparison of Cryptocurrencies," December 24, 2014. As of February 24, 2015: https://en.bitcoin.it/wiki/Comparison_of_cryptocurrencies

————, "Hardware Wallet," August 15, 2015a. As of February 19, 2015: https://en.bitcoin.it/wiki/Hardware_wallet

————, homepage, August 13, 2015b. As of February 25, 2015:
https://en.bitcoin.it/wiki/Main_Page

————, "Mining Hardware Comparison," September 16, 2015c. As June 25, 2015:
https://en.bitcoin.it/wiki/Mining_hardware_comparison

————, "Weaknesses," July 8, 2015d. As of February 16, 2015:
https://en.bitcoin.it/wiki/Weaknesses

Blanc, Jerome, "Thirty Years of Community and Complementary Currencies," *International Journal of Community Currency Research*, Vol. 16, 2012, pp. D1–4.

Blockchain, homepage, undated (a). As of June 25, 2015:
https://blockchain.info

————, "Market Capitalization," undated (b). As of February 19, 2015:
https://blockchain.info/charts/market-cap?timespan=all&showDataPoints
=false&daysAverageString=1&show_header=true&scale=0&address

————, "Send Via: Send Bitcoins Using Email and SMS," undated (c). As of February 19, 2015:
https://blockchain.info/wallet/send-via

Bonneau, Joseph, Andrew Miller, Jeremy Clark, Arvind Narayanan, Joshua A. Kroll, and Edward W. Felten, "Research Perspectives on Bitcoin and Second-Generation Cryptocurrencies," *Proceedings of IEEE Security and Privacy 2015*, San Jose, Calif.: IEEE Computer Society, May 2015.

Bonneau, Joseph, Arvind Narayanan, Andrew Miller, Jeremy Clark, and Joshua A. Kroll, "Mixcoin: Anonymity for Bitcoin with Accountable Mixes," *Financial Cryptography and Data Security: 18th International Conference*, Berlin: Springer Heidelberg, 2014, pp. 486–504.

Brantly, Aaron, "Financing Terror Bit by Bit," *CTC Sentinel*, Vol. 7, No. 10, October 2014, pp. 1–5.

Chaum, David, "Blind Signatures for Untraceable Payments," in David Chaum, Ronald L. Rivest, and Alan T. Sherman, eds., *Advances in Cryptology: Proceedings of Crypto '82*, Berlin: Springer-Verlag, 1983, pp. 199–203.

Chaum, David, Amos Fiat, and Moni Naor, "Untraceable Electronic Cash," in Shafi Goldwasser, ed., *Advances in Cryptology: Proceedings of Crypto '88: Proceedings*, Berlin: Springer-Verlag, 1990, pp. 319–327.

Christin, Nicolas, "Traveling the Silk Road: A Measurement Analysis of a Large Anonymous Online Marketplace," *Proceedings of the 22nd International Conference on World Wide Web (WWW 2013)*, Rio de Janeiro: World Wide Web Conference, 2013, pp. 213–223.

Cohen, Benjamin J., *The Geography of Money*, Ithaca, N.Y.: Cornell University Press, 1998.

CoinJoin, "Weaknesses in SharedCoin," undated. As of February 22, 2015:
http://www.coinjoinsudoku.com

CoinMarketCap, "Crypto-Currency Market Capitalizations," September 30,
2015a. As of June 25, 2015:
http://coinmarketcap.com

Covert, Adrian, "There's a Virus That Will Steal All Your Bitcoins," *Gizmodo.com*,
June 17, 2011. As of February 25, 2015:
http://gizmodo.com/5813039/theres-a-virus-that-will-steal-all-your-bitcoins

Danezis, George, Cédric Fournet, Markulf Kohlweiss, and Bryan Parno,
"Pinocchio Coin: Building Zerocoin from a Succinct Pairing-Based Proof System,"
*PETShop '13: Proceedings of the First ACM Workshop on Language Support for
Privacy-Enhancing Technologies*, New York: Association for Computing Machinery,
2013, pp. 27–30.

Daragahi, Borzo, "ISIS Declares Its Own Currency," *Financial Times* online,
November 13, 2014. As of February 24, 2015:
http://www.ft.com/intl/cms/s/2/baf893e0-6b4f-11e4-9337
-00144feabdc0.html#axzz3SgRLthZp

Dark Wallet, homepage, undated. As of February 22, 2015:
https://www.darkwallet.is

Dash, homepage, undated (a). As of June 25, 2015:
https://www.dashpay.io

———, "Masternodes and Proof of Service," undated (b). As of June 25, 2015:
https://www.dashpay.io/masternodes2

Dash Talk, "Reply to Kristov's Paper," self-published article, September 11, 2014.
As of February 22, 2015:
https://dashcointalk.org/threads/reply-to-kristovs-paper.2325

Davies, Glyn, *A History of Money: From Ancient Times to the Present Day*, Chicago:
University of Chicago Press, 2005.

Defense Advanced Research Projects Agency, "DARPA I2O Demo Day Featured
Programs," May 21, 2014. As of October 7, 2015:
http://www.darpa.mil/attachments/DARPAI2ODemoDay_ProgramDescriptions.pdf

Defense Science Board, Department of Defense, *Task Force Report: Resilient
Military Systems and the Advanced Cyber Threat,* January 2013. As of September
30, 2015:
http://www.acq.osd.mil/dsb/reports/ResilientMilitarySystems.CyberThreat.pdf

Department of Justice, U.S. Attorney's Office, Southern District of New York,
"Indictment and Supporting Documents: U.S. v. Liberty Reserve et al.," May 28,
2013. As of February 22, 2015:
http://www.justice.gov/usao/nys/pressreleases/May13/LibertyReserveetalDocuments.php

Desan, Christine, *Making Money: Coin, Currency, and the Coming of Capitalism*, Oxford: Oxford University Press, 2014.

Dowd, Kevin, "Contemporary Private Monetary Systems," self-published paper, August 2013. As of February 26, 2015:
http://www.kevindowd.org/app/download/8477462997/Contemporary+Private
+Monetary+Systems.pdf?t=1380159881

El Defrawy, Karim, and Joshua Lampkins, "Founding Digital Currency on Secure Computation," *CCS '14: Proceedings of ACM SIGSAC Conference on Computer and Communications Security*, March 2014, pp. 1–14.

Ensafi, Roya, Philipp Winter, Abdullah Mueen, and Jedidiah R. Crandall, "Large-Scale Spatiotemporal Characterization of Inconsistencies in the World's Largest Firewall," self-published paper, October 3, 2014. As of February 22, 2015:
http://arxiv.org/pdf/1410.0735.pdf

European Banking Authority, "EBA Opinion on 'Virtual Currencies,'" July 4, 2014. As of October 1, 2015:
https://www.eba.europa.eu/documents/10180/657547/EBA-Op-2014
-08+Opinion+on+Virtual+Currencies.pdf

European Central Bank, *Virtual Currency Schemes*, October 2012. As of October 1, 2015:
https://www.ecb.europa.eu/pub/pdf/other/virtualcurrencyschemes201210en.pdf

Eyal, Ittay, and Emin Gun Sirer, "Majority Is Not Enough: Bitcoin Mining Is Vulnerable," in Nicolas Christin and Reihaneh Safavi-Naini, eds., *Financial Cryptography and Data Security: 18th International Conference, FC 2014*, March 2014, pp. 436–454.

Federal Bureau of Investigation, "Ransomware on the Rise: FBI and Partners Working to Combat This Cyber Threat," January 20, 2015. As of February 13, 2015:
http://www.fbi.gov/news/stories/2015/january/ransomware-on-the-rise

Folding Coin, "Announcing Scotcoin," February 5, 2015. As of February 13, 2015:
http://foldingcoin.net/2015/01/announcing-scotcoin

Frieden, Jeffry A., *Global Capitalism: Its Fall and Rise in the Twentieth Century*, New York: W. W. Norton and Company, 2006.

Garay, Juan, Aggelos Kiayias, and Nikos Leonardos, "The Bitcoin Backbone Protocol: Analysis and Applications," in Elisabeth Oswald and Marc Fischlin, eds., *Advances in Cryptology—EUROCRYPT 2015: 34th Annual International Conference on the Theory and Applications of Cryptgraphic Techniques*, April 2015, pp. 281–310.

GHash.io, "Bitcoin Mining Pool GHash.IO Is Preventing Accumulation of 51 Percent of All Hashing Power," undated. As of February 23, 2015:
https://ghash.io/ghashio_press_release.pdf

Gill, Nathan, "Ecuador Turning to Virtual Currency After Oil Loans," *Bloomberg News* online, August 11, 2014. As of February 13, 2015:
http://www.bloomberg.com/news/articles/2014-08-11/
ecuador-turning-to-virtual-currency-after-oil-loans-correct-

GitHub, "Omni Protocol Specification (formerly Mastercoin)," undated. As of February 26, 2015:
https://github.com/OmniLayer/spec

Gomez, Georgina, "Sustainability of the Argentine Complementary Currency Systems: Four Governance Systems," *International Journal of Community Currency Research*, Vol. 16, 2012, pp. D80–89.

Helleiner, Eric, *The Making of National Money: Territorial Currencies in Historical Perspective*, Ithaca, N.Y.: Cornell University Press, 2003.

Hern, Alex, "Bitcoin Goes National with Scotcoin and Auroracoin," *Guardian* website, March 25, 2014. As of October 7, 2015:
http://www.theguardian.com/technology/2014/mar/25
/bitcoin-goes-national-with-scotcoin-auroracoin

Irish Coin, homepage, undated. As of February 24, 2015:
http://irishcoin.org

Ithaca Hours, homepage, undated. As of February 24, 2015:
http://www.ithacahours.com

Jack, William, and Tavneet Suri, "The Economics of M-Pesa," second version, self-published paper, August 2010. As of February 19, 2015:
http://www.mit.edu/~tavneet/M-PESA.pdf

Johnson, Marion, "The Cowrie Currencies of West Africa. Part I," *Journal of African History*, Vol. 11, No. 1, 1970, pp. 17–49.

Kaelberer, Matthias, "Trust in the Euro: Exploring the Governance of a Supra-National Currency," *European Societies*, Vol. 9, No. 4, 2007, pp. 623–642.

Kaspersky Labs, "The Desert Falcons Targeted Attacks," version 2.0, corporate publication, Moscow, 2015. As of October 1, 2015:
https://securelist.com/files/2015/02/The-Desert-Falcons-targeted-attacks.pdf

Kharif, Olga, "Bitcoin: Not Just for Libertarians and Anarchists Anymore," *BloombergBusiness.com*, October 9, 2014. As of February 20, 2015:
http://www.bloomberg.com/bw/articles/2014-10-09
/bitcoin-not-just-for-libertarians-and-anarchists-anymore

Kindleberger, Charles, *A Financial History of Western Europe*, Oxford: Oxford University Press, 1993.

King, Sunny, "Primecoin: Cryptocurrency with Prime Number Proof-of-Work," self-published paper, July 7, 2013. As of February 19, 2015:
http://primecoin.io/bin/primecoin-paper.pdf

King, Sunny, and Scott Nadal, "PPCoin: Peer-to-Peer Crypto-Currency with Proof-of-Stake," self-published paper, August 19, 2012. As of February 24, 2015: http://archive.org/stream/PPCoinPaper/ppcoin-paper_djvu.txt

Krebs, Brian, "True Goodbye: 'Using Truecrypt Is Not Secure,'" *KrebsonSecurity.com*, May 14, 2014. As of February 19, 2015: http://www.krebsonsecurity.com/2014/05/true-goodbye-using-truecrypt-is-not-secure/

———, "U.S. Government Seizes LibertyReserve.com," *KrebsonSecurity.com*, May 13, 2013. As of September 29, 2015: http://www.krebsonsecurity.com/2013/05/u-s-government-seizes-libertyreserve-com

Kroll, Joshua A., Ian C. Davey, and Edward W. Felten, "The Economics of Bitcoin Mining or, Bitcoin in the Presence of Adversaries," paper presented at the 12th Workshop on the Economics of Information Security (WEIS 2013), Washington, D.C., June 11–12, 2013.

Kuhn, Richard D., Vincent C. Hu, W. Timothy Polk, and Shu-Jen Chang, *Introduction to Public Key Technology and the Federal PKI Infrastructure*, Gaithersburg, Md.: National Institute of Standards and Technology, U.S. Department of Commerce, February 26, 2001. As of February 16, 2015: http://csrc.nist.gov/publications/nistpubs/800-32/sp800-32.pdf

Lajka, Arijeta, "Islamic State Takes a Stab at Legitimacy with Alleged Identification Cards as Forces Lose Ground in Iraq," *Vice News* online, April 16, 2015. As of June 25, 2015: https://news.vice.com/article/islamic-state-takes-a-stab-at-legitimacy-with-alleged -identification-cards-as-forces-lose-ground-in-iraq

Lee, Timothy B., "Major Glitch in Bitcoin Network Sparks Sell-Off; Price Temporarily Falls 23%," *Ars Technica*, March 11, 2013. As of April 16, 2015: http://arstechnica.com/business/2013/03/major-glitch-in-bitcoin-network-sparks -sell-off-price-temporarily-falls-23

Litecoin, homepage, undated. As of February 24, 2015: https://litecoin.org

Maidsafe, homepage, undated (a). As of on February 24, 2015: http://maidsafe.net

———, "SAFE Network System Docs," undated (b). As of February 24, 2015: http://systemdocs.maidsafe.net

Mas, Ignacio, and Dan Radcliffe, "Mobile Payments Go Viral: M-PESA in Kenya," World Bank website, March 2010. As of February 19, 2015: http://siteresources.worldbank.org/AFRICAEXT/Resources/258643 -1271798012256/M-PESA_Kenya.pdf

Mazacoin, homepage, undated. As of February 24, 2015: https://mazacoin.org

McMillan, Robert, "The Inside Story of Mt. Gox, Bitcoin's $460 Million Disaster," *Wired* online, March 3, 2014. As of September 29, 2015: http://www.wired.com/2014/03/bitcoin-exchange

Meiklejohn, Sarah, Marjori Pomarole, Grant Jordan, Kirill Levchenko, Damon McCoy, Geoffrey M. Voelker, and Stefan Savage, "A Fistful of Bitcoins: Characterizing Payments Among Men with No Names," *Proceedings of the 2013 Conference on Internet Measurement (IMC '13)*, October 2013, pp. 127–140.

Murphy, Edward V., M. Maureen Murphy, and Michael V. Seitzinger, *Bitcoin: Questions, Answers, and Analysis of Legal Issues*, Washington, D.C.: Congressional Research Service, August 14, 2015.

Nakamoto, Satoshi, "Bitcoin: A Peer-to-Peer Electronic Cash System," self-published paper, 2008. As of February 15, 2015: https://bitcoin.org/bitcoin.pdf

Namecoin, homepage, undated. As of February 24, 2015: http://namecoin.info/

Nxt Wiki, "Whitepaper:NXT," modified July 13, 2014. As of October 7, 2015: http://wiki.nxtcrypto.org/wiki/Whitepaper:Nxt

Only Coin, homepage, undated. As of February 19, 2015: https://onlycoin.com

Open Hub, "Bitcoin Project Summary," undated. As of February 26, 2015: https://www.openhub.net/p/bitcoin

Perfect Money, homepage, undated. As of April 16, 2015: https://perfectmoney.is

Pfajfar, Damjan, Giovanni Sgro, and Wolf Wagner, "Are Alternative Currencies a Substitute or a Complement to Fiat Money? Evidence from Cross-Country Data," *International Journal of Community Currency Research*, Vol. 16, 2012, pp. 45–56.

Pitta, Julie, "Requiem for a Bright Idea," *Forbes* online, November 1, 1999. As of February 26, 2015: http://www.forbes.com/forbes/1999/1101/6411390a.html

Pouwelse, Johan, Paweł Garbacki, Dick Epema, and Henk Sips, "The Bittorrent P2P File-Sharing System: Measurements and Analysis," in Miguel Castro, ed., *IPTPS 2005 Proceedings of the Fourth International Conference on Peer-to-Peer Systems*, February 2005, pp. 205–216.

Prisco, Giulio, "An Independent Scotland Powered by Bitcoin?" *CryptoCoinNews.com*, September 17, 2014. As of February 13, 2015: https://www.cryptocoinsnews.com/an-independent-scotland-powered-by-bitcoin

Recorded Future, "How Al-Qaeda Uses Encryption Post-Snowden (Part 1)," self-published paper, May 8, 2014a. As of February 17, 2015: https://www.recordedfuture.com/al-qaeda-encryption-technology-part-1

————, "How Al-Qaeda Uses Encryption Post-Snowden (Part 2)—New Analysis in Collaboration with ReversingLabs," self-published paper, August 1, 2014b. As of February 17, 2015:
https://www.recordedfuture.com/al-qaeda-encryption-technology-part-2

Ripple, "FAQ," undated (a). As of February 22, 2015:
http://wiki.ripple.com/FAQ

————, homepage, undated (b). As of February 24, 2015:
https://ripple.com

Rogoff, Kenneth, "Costs and Benefits to Phasing Out Paper Currency," *NBER Macroeconomics Annual 2014*, Vol. 29, 2015, pp. 445–456.

Salt Spring Dollars, homepage, undated. As of February 24, 2015:
http://www.saltspringdollars.com

Samani, Raj, "Cybercrime Exposed: Cybercrime-as-a-Service," corporate white paper, Santa Clara, Calif.: McAfee Labs, 2013a. As of September 29, 2015:
http://www.mcafee.com/us/resources/white-papers/wp-cybercrime-exposed.pdf

————, "Digital Laundry: An Analysis of Online Currencies, and Their Use in Cybercrime," corporate white paper, Santa Clara, Calif.: McAfee Labs, 2013b. As of September 29, 2015:
http://www.mcafee.com/us/resources/white-papers/wp-digital-laundry.pdf

Saroiu, Stefan, P. Krishna Gummadi, and Steven D. Gribble, "A Measurement Study of Peer-to-Peer File Sharing Systems," Martin G. Kienzle and Prashant J. Shenoy, eds., *Proceedings of SPIE: Multimedia Computing and Networking (MMCN) 2002,* Vol. 4673, 2002, pp. 156–170.

Scotcoin, homepage, undated. As of February 19, 2015:
http://scotcoin.org

Square, homepage, undated. As of February 19, 2015:
https://squareup.com

Taylor, Adam, "The Islamic State (or Someone Pretending to Be It) Is Trying to Raise Funds Using Bitcoin," *Washington Post* online, June 9, 2015. As of June 25, 2015:
http://www.washingtonpost.com/blogs/worldviews/wp/2015/06/09/the-islamic
-state-or-someone-pretending-to-be-it-is-trying-to-raise-funds-using-bitcoin

Taylor, Michael Bedford, "Bitcoin and the Age of Bespoke Silicon," paper presented at the *International Conference on Compilers, Architecture, and Synthesis for Embedded Systems (CASES)*, Montreal, Quebec, September 29–October 4, 2013.

Tor Project, "Anonymity Online," undated (a). As of February 16, 2015:
https://www.torproject.org

———, "Category, Tags, Attacks," December 19, 2014. As of February 24, 2015:
https://blog.torproject.org/category/tags/attacks

———, "Hidden Service Protocol," undated (b). As of February 24, 2015:
https://www.torproject.org/docs/hidden-services.html.en

———, homepage, undated (c). As of February 22, 2015:
https://www.torproject.org

———, "Overview," undated (d). As of February 24, 2015:
https://www.torproject.org/about/overview

———, "Security Advisory Relay Early Traffic," July 30, 2014. As of February 24,
2015:
https://blog.torproject.org/blog/tor-security-advisory-relay-early-traffic-confirmation-attack

Totnes Pound, homepage, undated. As of February 24, 2015:
http://www.totnespound.org

Treisman, Daniel, "Russia's 'Ethical Revival': The Separatist Activism of Regional
Leaders in a Postcommunist Order," *World Politics*, Vol. 49, No. 2, 1997,
pp. 212–249.

United States v. Liberty Reserve, 13 CRIM368 (S.D.N.Y. 2013). As of February 22,
2015:
http://www.justice.gov/usao/nys/pressreleases/May13/LibertyReservePR
/Liberty%20Reserve,%20et%20al.%20Indictment%20-%20Redacted.pdf

Vandervort, David, Dale Gaucas, and Robert St. Jacques, "Issues in Designing a
Bitcoin-Like Community Currency," paper presented at the Second Workshop on
Bitcoin Research, San Juan, Puerto Rico, January 30, 2015.

Warren, Jonathan, "Bitmessage: A Peer-to-Peer Message Authentication and
Delivery System," self-published paper, November 27, 2012. As of February 23,
2015:
https://bitmessage.org/bitmessage.pdf

Weatherford, Jack McIver, *The History of Money*, New York: Crown Publishers,
1997.

WebMoney Transfer, homepage, undated. As of April 16, 2015:
http://www.wmtransfer.com

Wikipedia, "Ora (Currency)," April 27, 2015. As of April 21, 2015:
http://en.wikipedia.org/wiki/Ora_%28currency%29

Willett, J. R., *The Second Bitcoin White Paper*, vs. 0.5 (Draft for Public Comment),
self-published paper, undated. As of October 1, 2015:
https://sites.google.com/site/2ndbtcwpaper/2ndBitcoinWhitepaper.pdf

Winter, Philipp, and Stefan Lindskog, "How the Great Firewall of China is
Blocking Tor," paper presented at the Second USENIX Workshop on Free and
Open Communication on the Internet (FOCI), Bellevue, Wash., August 2012.

World Bank, *Global Findex Database*, 2014. As of June 25, 2015:
http://datatopics.worldbank.org/financialinclusion

Zerocash Project, homepage, undated. As of February 22, 2015:
http://zerocash-project.org

Zerocoin Project, homepage, undated. As of February 22, 2015:
http://zerocoin.org